Above: The first trees on Earth, e.g. tree ferns, giant horsetails and club mosses, appeared in the Carboniferous period, c.330 million years ago, 100 million years before the dinosaurs.

Published by Wooden Books LLC,
San Rafael, California

Hardcover edition originally published
in the US as *The Miracle of Trees* in 2012
by Bloomsbury Publishing USA, New York

Library of Congress Cataloging-in-Publication Data
Huikari, O.
Trees, and how they work

Library of Congress Cataloging-in-Publication
Data has been applied for

ISBN-10: 1-952178-11-8
ISBN-13: 978-1-952178-11-5

Designed and typeset in Glastonbury, UK

Printed in China on 100% FSC
approved sustainable papers by FSC
RR Donnelley Asia Printing Solutions Ltd.

TREES

AND HOW THEY WORK

Olavi Huikari

Translated from the Finnish by Matti Pohjonen, with the assistance of FILI, the Finnish Literature Exchange. Many thanks to the brilliant librarians at Helsinki University Library and Finnish Museum of Natural History.

Engravings throughout from: *Der Wöld* by E. A. Rokmakler, Leipzig & Heidelberg, 1863; *Morpologie der Gewächse* by W. Hofmeister, Leipzig, 1868; *Verggleichende Anatomie der Vegetationsorgane der Phanerogamen und Farne* by A. de Bary, Leipzig, 1877; *Pflanzenleben* by A. K. von Marilaun, Leipzig & Wien, 1900. Woodcuts on pages 19, 30 and 43 by Gwen Raverat. Drawings on pages 3–4, 9, 11, 15, 18, 20, 21, 22, 24, 26–29, 36–37 & 57 by Vivien Martineau; 6–7, 12, 16–17, 25, 43 & 52 by Matt Tweed. Further reading, and thanks to: *Trees, Their Natural History* by Peter Thomas, 2000; *The Secret Life of Trees* by Colin Tudge, 2006.

CONTENTS

Introduction

THE BOOK YOU ARE HOLDING in your hands is the outcome of many years of scientific research into the lives of trees. It gives a glimpse into their mysterious world, a world we as humans have always been connected to as intrinsic parts of nature, but know very little about.

A group of trees forms a forest and a forest offers food and protection to sustain life's great diversity. One individual tree can live hundreds of years, closely collaborating with other species of trees, plants, micro-organisms and animals, and it is this which gives a tree its miraculous power to flourish, even under the most extreme conditions. Trees are breadbaskets to many, but also resilient fighters, steadfastly protecting their own ground.

The Finnish epic *Kalevala* tells us how important trees are. When the hero Väinämöinen swims to a barren world, he asks the Great Bear for help, and a boy, Pellervo, is sent, who begins planting trees.

Scientists today are able to simulate the environments and conditions in which trees live, and in these meticulous experiments trees have been given a voice to express themselves. By studying their various growth patterns we have been able to listen, learn and discover the important things in a tree's life—what it likes and dislikes, what helps it grow, and, finally, what kills it. We cannot conserve what we do not understand.

Years of persistent research into plant physiology and forest ecology have provided some answers to the mystery of trees. The aim of this book is therefore to help you better understand the miracle of a tree, its parts, purposes and dynamic changes, and the way its life is interwoven into the very definition of what it means to be human.

What Is a Tree?

getting to know your cousins

Trees are plants which have learned how to grow very tall, using a self-supporting perennial woody stem. Like all life on Earth they descend from archæbacteria, primarily creative photosynthesizers and scavenging methane producers, which long ago began to regulate the planetary environment. About 3 billion years ago these early single-celled lifeforms diversified into new breeds of mobile consumers, some of whom incorporated other single-celled organisms, to mutual benefit. From here on, a succession of such endosymbiotic relationships generated a panoply of protoctists, and different streams of these, with characteristic capabilities, became the three multicellular kingdoms: Fungi, Animalia and Plantæ (*shown in the diagram opposite, after Lynn Margulis*).

The first vascular plants appeared in the Silurian period, 400 million years ago (mya), and began to pump oxygen into the atmosphere. By the time of the Carboniferous (around 330mya), carbon dioxide was getting scarce and tree ferns evolved with leaves that breathed. Next came height, with the gymnosperms—the ginkgos and towering conifers of the late Permian (250mya) and Jurassic (150mya). Then, finally, angiosperms, or flowering plants, in the form of hardwood trees like magnolias, overtook conifers around 75mya.

Trees are particularly characterized by their manufacture of *lignin* as a secondary cell wall (around 30% of the dry mass of wood), and they also, like all plants, produce vast quantities of *cellulose* (the most common organic compound on Earth) as well as *tannins*.

We humans share about 50% of our DNA with trees—something to think about next time you sit under an old oak!

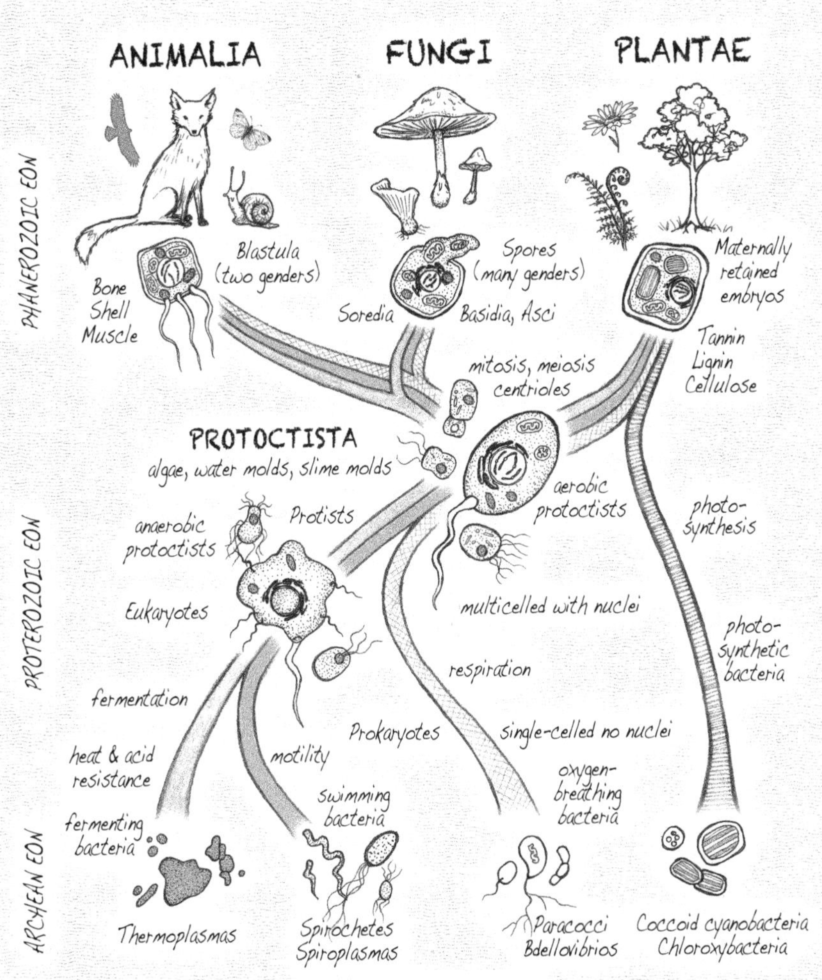
ANIMALIA
FUNGI
PLANTAE
PHANEROZOIC EON
Blastula
(two genders)
Bone
Shell
Muscle
Spores
(many genders)
Soredia
Basidia, Asci
Maternally
retained
embryos
Tannin
Lignin
Cellulose
mitosis, meiosis
centrioles
PROTOCTISTA
algae, water molds, slime molds
aerobic
protoctists
PROTEROZOIC EON
anaerobic
protoctists
Protists
photo-
synthesis
Eukaryotes
multicelled with nuclei
photo-
synthetic
bacteria
respiration
fermentation
Prokaryotes
single-celled no nuclei
heat & acid
resistance
motility
oxygen-
breathing
bacteria
swimming
bacteria
fermenting
bacteria
ARCHEAN EON
Thermoplasmas
Spirochetes
Spiroplasmas
Paracocci
Bdellovibrios
Coccoid cyanobacteria
Chloroxybacteria
BACTERIA

Light Is the Fuel of Life

the sweet art of sugars

Light is the fuel of life. It powers the biochemical processes needed for photosynthesis, where trees and other plants combine water (H_2O) and carbon dioxide (CO_2) to produce sugars and oxygen.

In the leaves of plants, light energy from red and blue photons is captured in chloroplasts and used to separate protons from electrons in water molecules across the thylakoid membrane. In a series of reactions the energy is converted into *adenosine diphosphate* (ATD) and *nicotinamide adenine dinucleotide phosphate* (NADPH), releasing oxygen as a byproduct. The carbon liberated from the CO_2 is then fixed into the skeletons of small sugars like glucose and fructose, which are combined into larger structural sugars, like cellulose or lignin, or stored as energy reserves in the starches of roots, tubers, and seeds.

At night, trees make a huge range of phytochemicals using the energy harnessed during the day, transforming the small sugars into amino acids with help from soluble nitrogen from the roots, and then into medicinal or toxic alkaloids, related phenolics like vanilloids, salicylates, and balsams, and polymers like astringent tannins and woody lignin. Plants can also reduce sugars to lipids to make the edible oils of seed energy stores, protective soaps, and aromatic oils like menthol, limonene, and camphor, and resins and latexes like myrrh and rubber.

LIFE NEEDS PLANTS

understanding food chains

Apart from a very few organisms who draw energy from hot oceanic vents, all life on Earth depends either directly or indirectly on energy captured from the Sun. At the bottom of the food chain are plants, *autotrophs*, which convert carbon dioxide and water into chemicals, such as glucose, which store the sun's energy. The biggest plants are trees. Above autotrophs are *consumers*, or *heterotrophs*, who get their energy from eating autotrophs or other heterotrophs. Food chains are therefore often organized by scientists in trophic levels (*below and opposite top left*). Some consumers, like bacteria and fungi, are *decomposers*, and get their energy by chemically reacting with dead things which may have already been chewed by *detrivores*, like snails or vultures.

All living organisms, animals and plants, have tiny battery-like *mitochondria* in their cells which *respire*, so breathe out water and CO_2 (the opposite to photosynthesis), as they maintain key processes, such as making ATP. ATP is the chemical energy used *inside* cells, and the human body turns over its own weight in ATP every day. The net productivity of any system is its gross productivity less its respiratory loss (*e.g., leaf example opposite lower left, after Rutherford*). Apart from the cycling of carbon in photosynthesis and respiration, other essential biogeochemical cycles involve water, nitrogen, sulphur (in proteins and enzymes), phosphorus (in DNA), and other trace minerals.

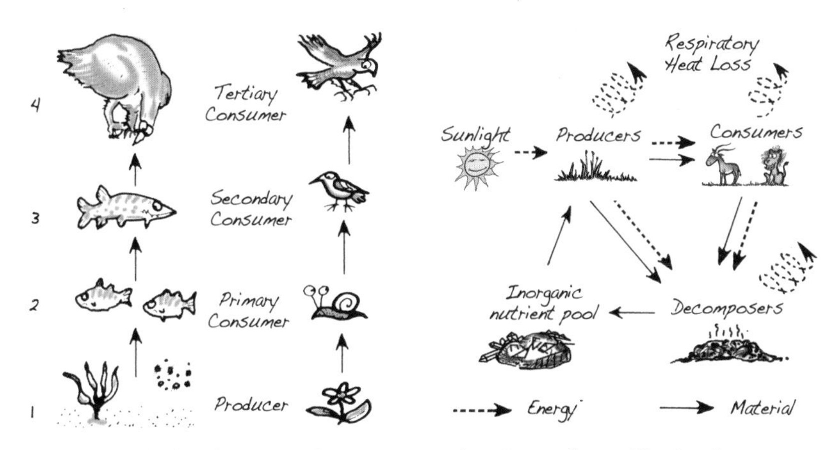

Above: Four trophic levels. Producers capture solar energy. Vegetarian primary consumers are eaten by secondary consumers. Only 10% of the captured energy is transmitted to the next level, so tertiary consumers are rare.

Above: Energy and material flow through an ecosystem. Decomposers help recycle nutrients back to primary producers. Soil respiration (caused by bacteria in soil) is an important factor in the global carbon cycle.

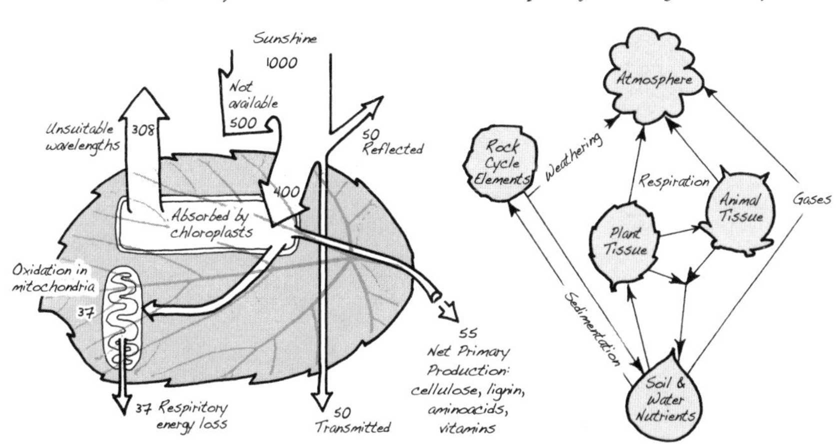

Above: Chloroplasts make glucose from red and blue light, while mitochondria burn glucose to power cells, causing respiration. From an input of 1,000 units of energy, the gross primary productivity is 92 (37 + 55).

Above: Biogeochemical cycles. 40 vital elements cycle through ecosystems. The reservoir for these is in rocks and soil, delivered through weathering, sedimentation and uptake by plants or animals licking rocks.

How Many Trees Are There?

a habit of reaching for the stars

You might think trees are only found in one or two families of "tree-like plants," but this is not the case. The diagram opposite shows the family tree of all plants, with dots beside orders which contain significant trees. Evidently, most orders have discovered the tree "form"—it works well, evidence of convergent evolution.

Scientists today estimate that there are about 60,000 different species of flowering trees, one fifth of the 300,000 species of flowering plants (angiosperms). Of the more ancient lineages of plants which used to rule the Earth, there remain around 800 species of tree-ferns, but alas none of the 100ft giant horsetails of the Paleozoic, and of the gymnosperms only 130 species of palm-like cycads, just one lonely species of ginkgo, and 630 species of conifers (*see too pages 54–55*).

Conifers include the tallest plants on Earth, the coastal redwoods, which can grow up to 380 feet. They are one of only two remaining species of sequoias which long ago ruled the world but have now largely vanished. Pine, birch and other deciduous trees arrived in northern latitudes only after the last Ice Age, around 10,000 years ago.

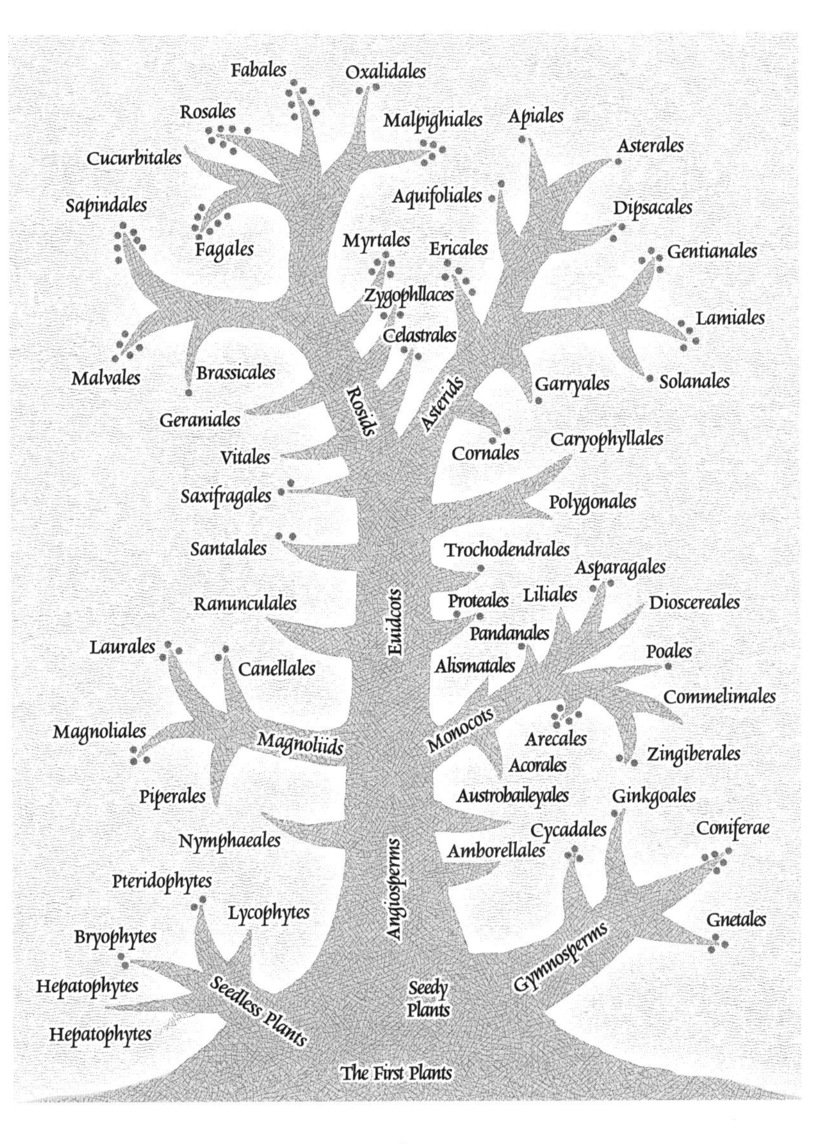
Fabales
Oxalidales
Rosales
Malpighiales
Apiales
Cucurbitales
Asterales
Aquifoliales
Sapindales
Dipsacales
Fagales
Myrtales
Ericales
Gentianales
Zygophllaces
Lamiales
Celastrales
Malvales
Brassicales
Solanales
Garryales
Rosids
Asterids
Geraniales
Caryophyllales
Cornales
Vitales
Saxifragales
Polygonales
Santalales
Trochodendrales
Asparagales
Proteales
Liliales
Dioscereales
Ranunculales
Euidcots
Pandanales
Laurales
Poales
Alismatales
Canellales
Commelimales
Magnoliales
Magnoliids
Monocots
Arecales
Zingiberales
Acorales
Piperales
Austrobaileyales
Ginkgoales
Cycadales
Coniferae
Nymphaeales
Amborellales
Pteridophytes
Lycophytes
Angiosperms
Gnetales
Bryophytes
Gymnosperms
Hepatophytes
Seedless Plants
Seedy Plants
Hepatophytes
The First Plants

The Parts of a Tree

and how they fit together

The ciliated protists from which plants evolved propelled themselves between the aquatic sediment, where they acquired minerals, and the surface, where they absorbed sunlight. For large terrestrial plants like trees, accessing these resources from a single site is the primary concern. As baby plants grow from their primordial *meristem*, some cells push upward toward the light, others move with gravity underground, and between the two *cambium* cells differentiate into vascular *xylem* (wood) and *phloem* (bark), connecting the extremities, with storage cells at the core. Modules of root, stem, and leaf then branch off at characteristic frequencies, with variations such as tactile tendrils, protective thorns, and flower buds.

The wisdom of a tree is that its key production is decentralized into numerous units that operate independently at the molecular level, units so tiny that production can continue even if part of a leaf is destroyed or deprived of water or nutrients. Unlike most animals, the survival of the whole is not invested in any one module, allowing for generous donations of tissue, without fatality, to passing herbivores or parasites. Because of their holistic molecular unity, trees always "know" where to increase or decrease production and how the necessary ingredients should be distributed. As long as some meristem persists, entire trees can regenerate from a stump.

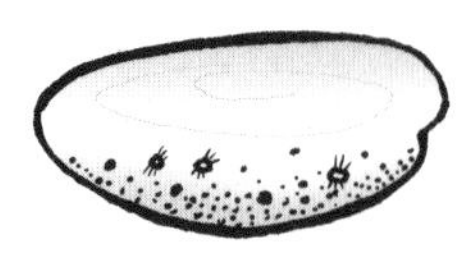

Stage 1. Algae & Micro-organisms build sediment.

Stage 2. Early Colonisers begin to grow in sediment.

Stage 3. Climax vegetation within healthy ecosystem.

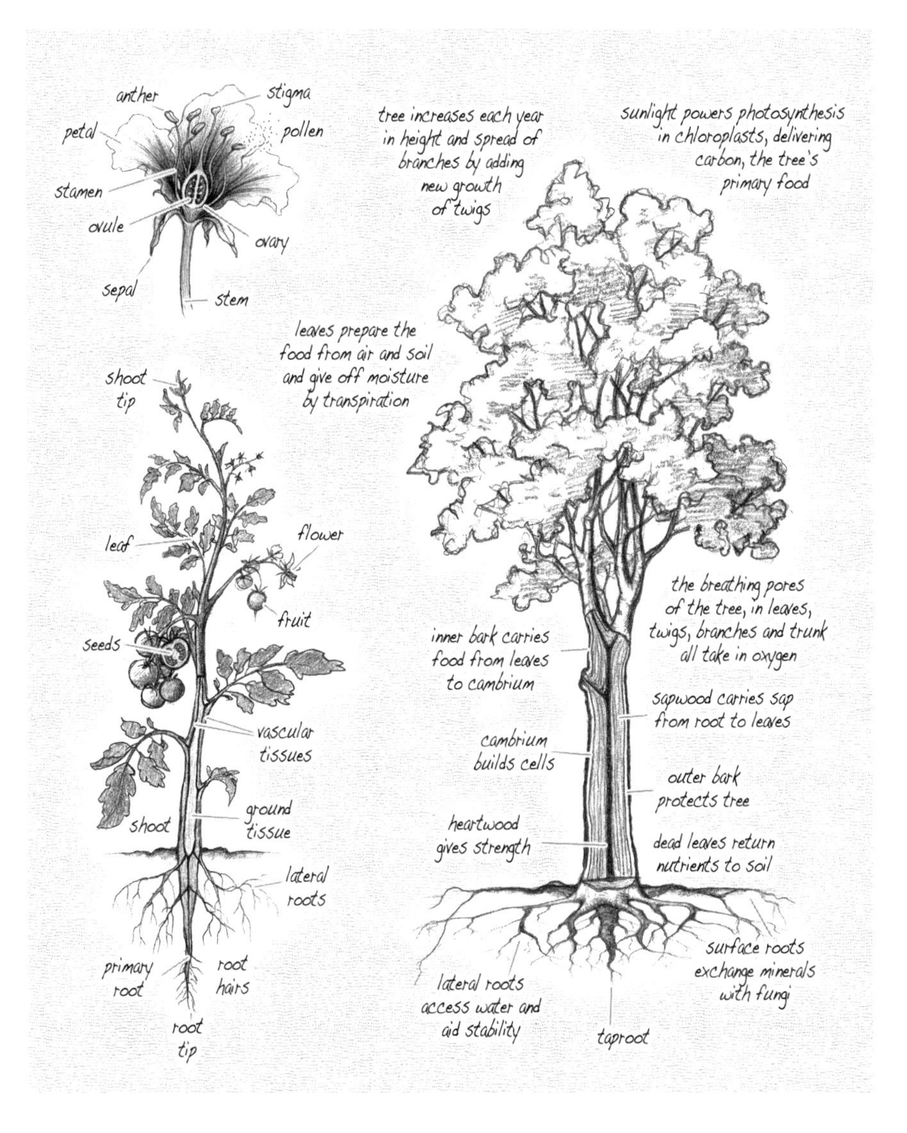
anther
stigma
petal
pollen
stamen
ovule
ovary
sepal
stem
tree increases each year in height and spread of branches by adding new growth of twigs
sunlight powers photosynthesis in chloroplasts, delivering carbon, the tree's primary food
leaves prepare the food from air and soil and give off moisture by transpiration
shoot tip
leaf
flower
fruit
seeds
vascular tissues
shoot
ground tissue
lateral roots
primary root
root hairs
root tip
the breathing pores of the tree, in leaves, twigs, branches and trunk all take in oxygen
inner bark carries food from leaves to cambium
sapwood carries sap from root to leaves
cambium builds cells
outer bark protects tree
heartwood gives strength
dead leaves return nutrients to soil
surface roots exchange minerals with fungi
lateral roots access water and aid stability
taproot

THE MIRACLE OF A TREE

and some important molecules

The miracle of a tree is how it transforms light into chemical energy. To begin this process, for photosynthesis to begin, the sun must first warm the cells. The reaction which occurs is then: *solar energy* + $6H_2O$ + $6CO_2$ = $C_6H_{12}O_6$ + $6O_2$ (*below*). On a sunny day a 100-year old beech tree breathes in around 35,000 litres of air, from which it extracts about 10,000 litres or 18kg of CO_2 to produce 12kg of sugar and 13kg of oxygen. An oak can have a quarter of a million leaves, and walking on a forest on a sunny day, you can literally feel the oxygen in the air. A tree's long life guarantees biodiversity and large renewable energy resources for all of life to use—all because of light.

Stored sugars are vital in deciduous trees, which need to survive a cold winter and then still have enough energy to firstly flower in the spring and then produce a full crown of leaves before photosynthesis begins once more. Energy is also stored in seeds.

In nature, the competition for light is fierce. A tree has to compete with other plant species for a place in the sun, and the one that gets sunlight thrives, while others shrivel in the shadows. With its sturdy trunk, a tree elevates its canopy above the competition, yet is still only able to photosynthesize about 1% of the light it receives, and of this only 10% is then converted into wood. A tree thus stores about a thousandth of the light energy that falls on it.

Above: Walnut trees flower as female clusters (which later ripen into fruit and nuts), and male catkins.

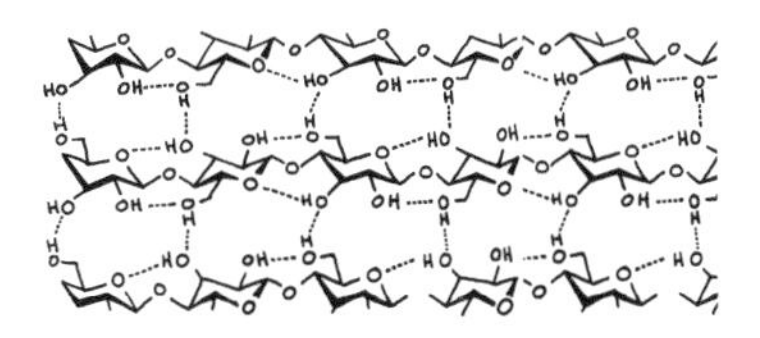

Above: Cellulose, a third of all plant matter, is composed of thousands of linked glucose units.

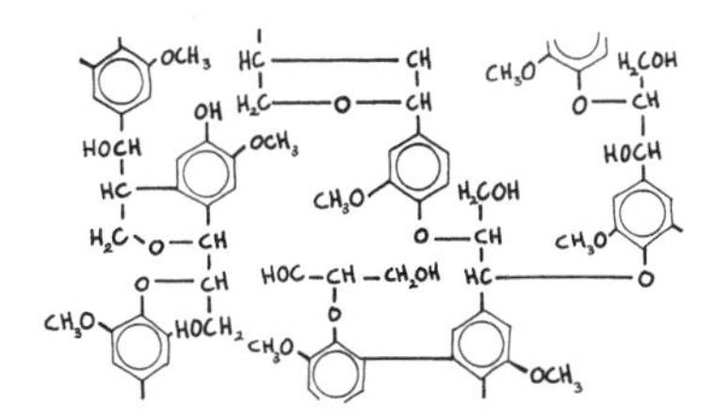

Above: The structure of woody lignin (detail), a complex compound, also a third of all plant matter.

Above: Larch, a deciduous conifer, common in the huge boreal forests of Russia and Canada.

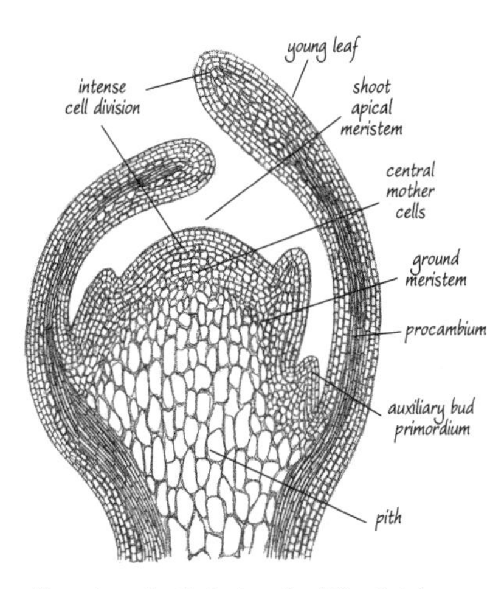

Above: A growing tip. Regions of undifferentiated or "meristematic" cells are where plant growth occurs.

Inside a Leaf

the green revolution

The leaf of a tree is composed of an outer waxy layer and *epidermic* cells that protect the interior cells where the vital photosynthesis takes place. Various pigments also protect these outer cells from damaging ultraviolet solar radiation, and in hot countries some trees alter their leaf angles to further reduce sunburn (eucalyptus leaves hang downward to escape the midday sun). Varying proportions of cell structures also result in different sun leaves and shade leaves.

Beneath the epidermis, located to receive the most sunlight, is the *parenchyma* tissue. Most photosynthesis takes place here, inside the factory halls of green *chloroplasts* in parenchyma cells.

Chloroplasts, like mitochondria, have their own circular DNA, and consist of two phospholipid membranes which enclose an aqueous fluid called *stroma*, in which stacks of small green thylakoid compartments do the work of photosynthesis. Chloroplasts are connected by *stromules*, and seem to function as a network. With a staggering total surface area of 350 square km in a large deciduous tree, they use about 95% of the red and blue wavelengths of light.

At the bottom of a leaf are microscopic air pores called *stomata*, which provide the cell with air and the carbon dioxide necessary for photosynthesis. There are about 100 to 700 stomata in each square millimeter, opening or closing depending on the amount of light.

Vascular bundles transport the water up from the roots to the cells for photosynthesis and then carry the sugary outputs back out from the leaf into the plant. Evaporating water enables the flow of new water (and its nutrients) to be distributed evenly throughout the tree.

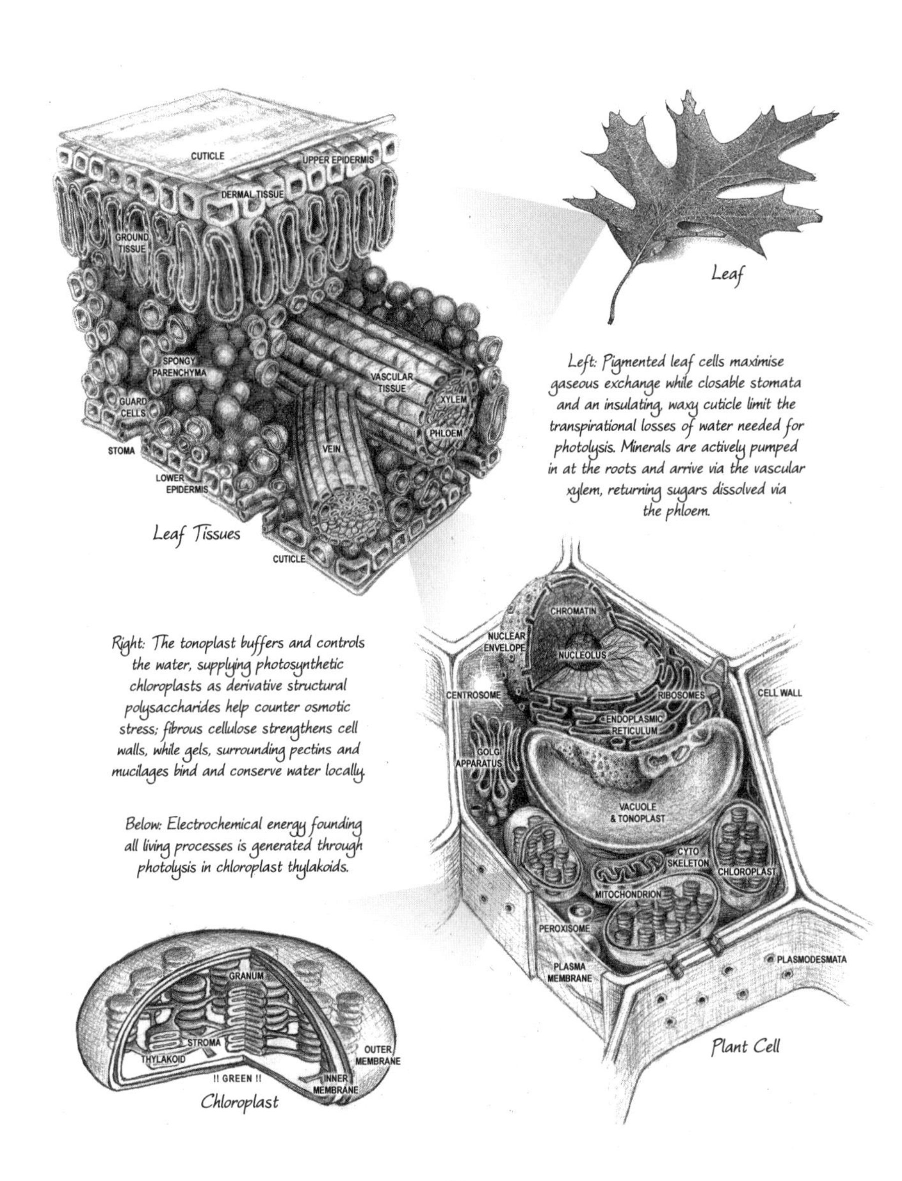

Leaf

Leaf Tissues

Left: Pigmented leaf cells maximise gaseous exchange while closable stomata and an insulating, waxy cuticle limit the transpirational losses of water needed for photolysis. Minerals are actively pumped in at the roots and arrive via the vascular xylem, returning sugars dissolved via the phloem.

Right: The tonoplast buffers and controls the water, supplying photosynthetic chloroplasts as derivative structural polysaccharides help counter osmotic stress; fibrous cellulose strengthens cell walls, while gels, surrounding pectins and mucilages bind and conserve water locally.

Below: Electrochemical energy founding all living processes is generated through photolysis in chloroplast thylakoids.

Plant Cell

Chloroplast

Evergreen or Deciduous

fight or flight

Trees living in year-round pleasant conditions, like wet tropical areas, tend to be as evergreen as they can, gradually exchanging old leaves for new ones, or occasionally shedding them all at once to replace them immediately. In difficult climates, however, where there are periods unfavorable for growth, many trees shed their leaves, resulting in bare branches for months during cold winters or hot dry summers. Toward the poles, to make use of the short growing season, evergreens once more appear, but in the very coldest and upper alpine regions deciduous trees once more predominate.

In all trees new leaves or needles are grown early, before the stem itself begins to grow in diameter. This happens differently in deciduous and coniferous trees, as deciduous trees have to wait until their leaves regrow every year, while coniferous trees only shed three- to five-year-old needles (*see cross-section of needle on page 57*).

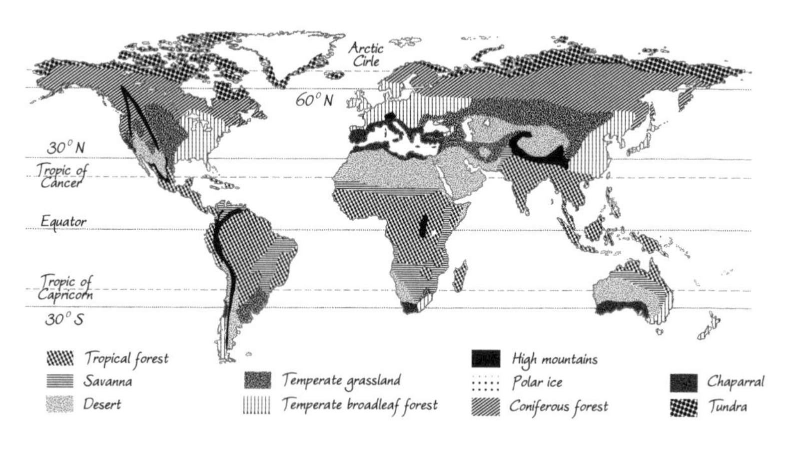

Above left: EVERGREEN *trees. All trees absorb vital nitrogen from the ground as ammonium and, with the help of a special enzyme, change it into glutamine and asparagine. Before winter, nitrogen is changed into argine and stored (especially in needles). Above right: In* DECIDUOUS *trees nitrogen is stored in buds and seeds, which also contain plentiful sugar which ferments into alcohol. Alcoholic rowanberries can get birds quite intoxicated, helping seeds get spread.*

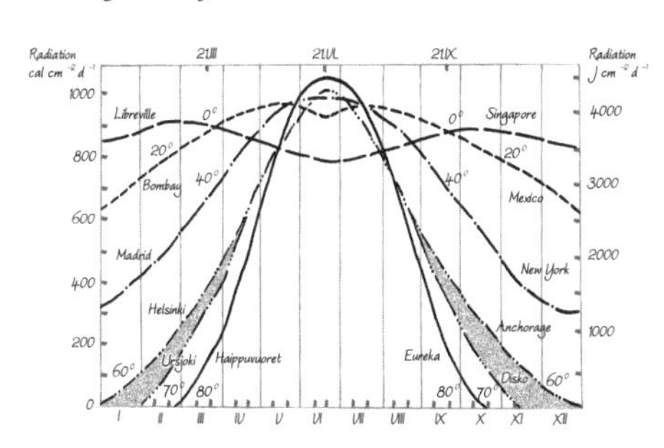

Above: The amount of sunlight falling per day at various latitudes in the different months of the year, highlighting the difference between the equator and poles.

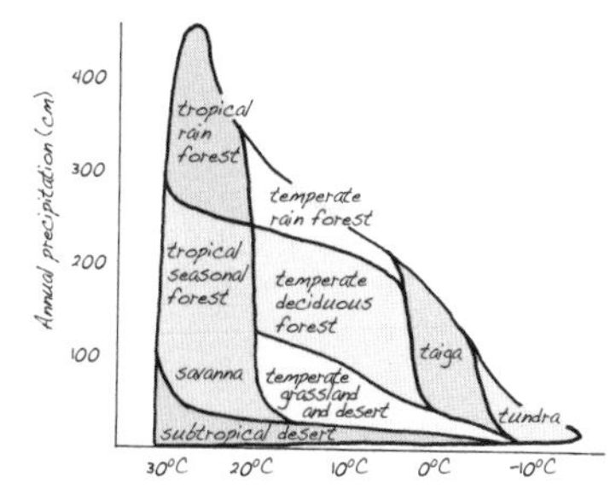

Above: The major biomes plotted against annual precipitation and temperature. Similar types of ecosystems may be found in similar biomes across the world.

THE COLORS OF A TREE

in black and white

The fantastic colors of autumn allow us to witness the different frequencies of light used by trees. When they are working, chloroplasts use light's red and blue wavelengths (around 700 and 450 nm respectively) to maintain a tree's life. The unused green light (600 nm) is then reflected to our eyes as we reflect on nature's beauty. Green means the tree is alive, and sucking in red and blue energy.

Drought can force trees to shrink their foliage. When this happens, enzymes begin breaking up the green *chlorophyll* particles, reducing their use of red light and reflecting it instead. Red and green make yellow, so we get the color of hot dry summers, and early autumn. Later, we also see another pigment that was there all along, *carotenoid*, with its characteristic yellows, oranges, and browns.

Autumn leaves can still produce lots of sugar, but cool night temperatures can prevent the sugar sap flowing down through the leaf veins. Trees, therefore, carefully remove all useful substances from their old leaves with *anthocyanin*, which gives autumn leaves their brilliant shades of red and purple. Only then does the base of the leafstalk grow cork cells to isolate each leaf from the rest of the tree with water-absorbent cells that break down easily when frozen, causing the leaves or needles to fall to the ground with the first frost.

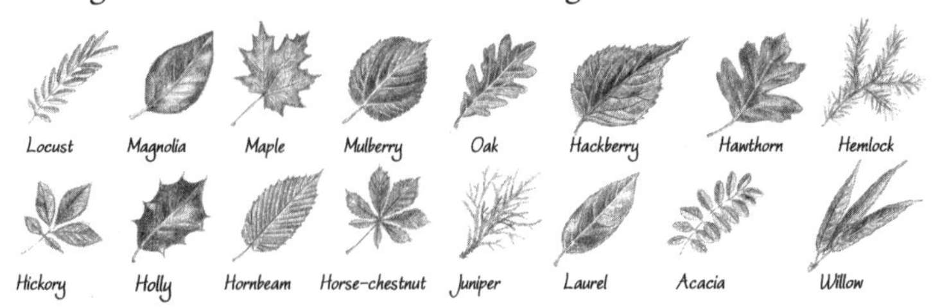

Above: Wood engraving by Gwen Raverat, 1936. Below: An autumn leaf changing color. As the chlorophyll particles are broken up the green color increasingly disappears from the leaves to produce reds and oranges.

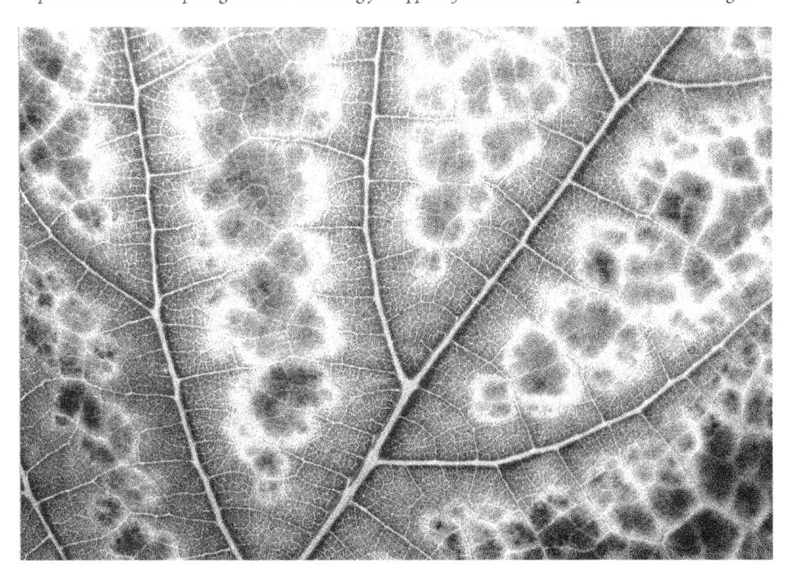

The Cambium

where it all happens

A tree's growth comes from its living meristematic tissue. Through repetitive cell divisions, an egg cell grows into an embryo. After a series of further cell divisions, specialization of different cells and formation of tree structures, we finally get what we call a tree.

The meristematic tissue of a tree exists everywhere between the bark and wood of the tree, and in the growth points at the branch and root ends. A living layer, sometimes only a single cell thick, it is like a thin sheet of film covering the entire tree.

The meristematic tissue responsible for widening a tree is called the *cambium*. Cells that the cambium form *inward*, toward the tree's center, become its *xylem*, or wood, which transports water up the tree; cells that form *outward* become the *phloem*, the transport channels for the material created in photosynthesis, and the outer bark. In some trees (*below left, after Peter Thomas*), wood cells grown in the spring, when the tree may drink huge amounts of water, have wider, lighter-colored vessels than cells grown later in the season, these differences producing the characteristic "rings" of a tree.

The meristematic tissue responsible for the height growth of a tree is found at the top of the stem, and at the ends of the roots, branches, and buds. These are called the "growth points" of a tree.

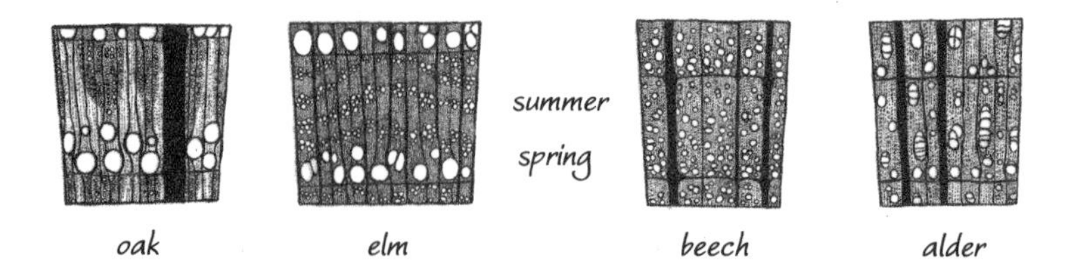

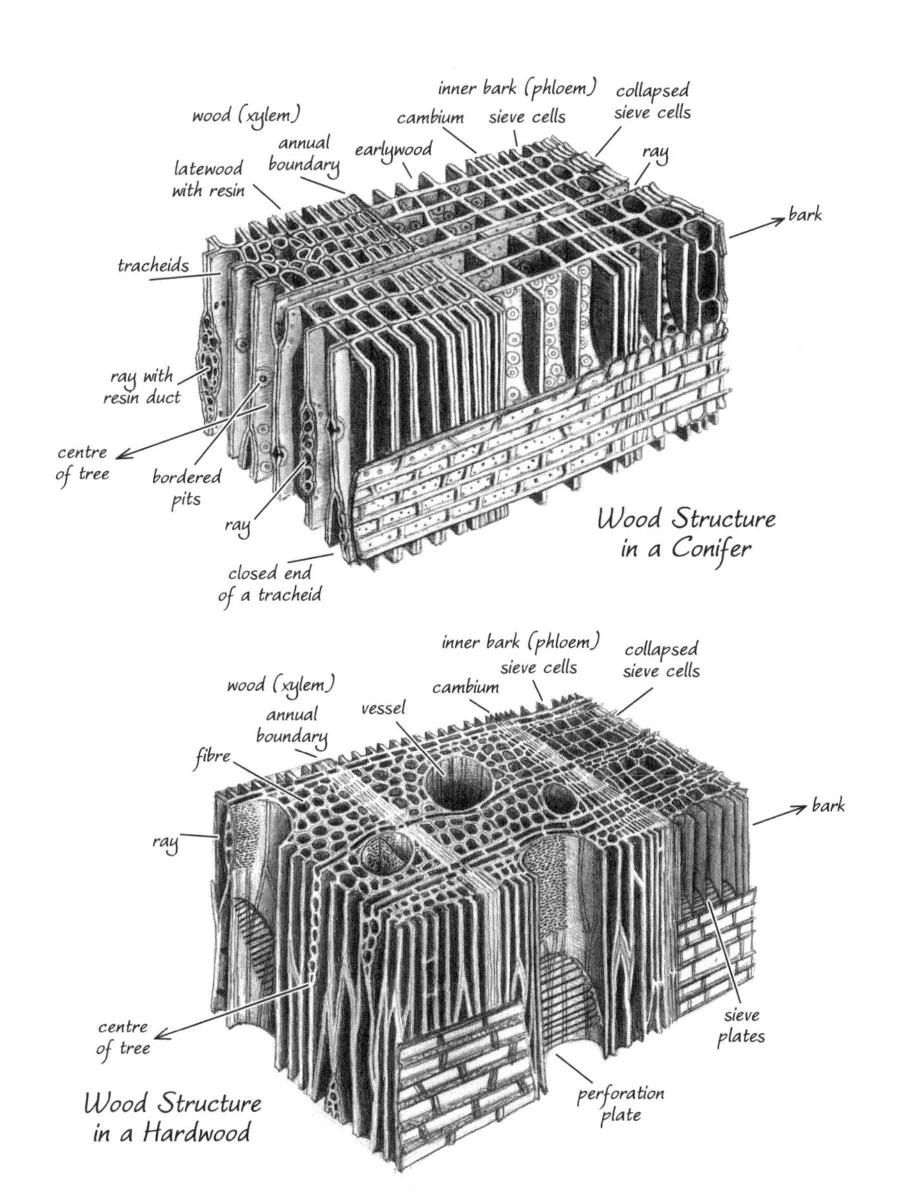

inner bark (phloem)
collapsed sieve cells
wood (xylem)
cambium
sieve cells
annual boundary
earlywood
ray
latewood with resin
bark
tracheids
ray with resin duct
centre of tree
bordered pits
ray
closed end of a tracheid
Wood Structure in a Conifer
inner bark (phloem)
collapsed sieve cells
sieve cells
wood (xylem)
cambium
annual boundary
vessel
fibre
bark
ray
centre of tree
sieve plates
perforation plate
Wood Structure in a Hardwood

THE BARK OF A TREE

safe and sound

Bark, the thick durable skin that protects the living parts of a tree against parasites and external damage, is formed as the thin living sheet of cambium in a tree creates the inner bark, or *phloem*, outward. The living part of the phloem consists of an array of sieve cells stacked up to form tubes through which the sap of the tree flows toward low pressure zones, upward, sideways or downward (at 10 cm per hour in larch and over 100cm per hour in ash). Unlike the inner woody xylem tubes, sieve tubes are not thickened with lignin so are thin-walled, and set in a mesh of strong fibers and rays.

The phloem, in turn, creates cork tissue outward, whose outer layer, outer bark, then cracks as the tree grows in diameter every year. Bark insulates against heat, and so, being both dry and water-resistant, is highly resistant to changes in the weather. It also prevents evaporation of liquids from the stem, something it does so well that it is used to stop wine bottles. However, most cork tissue possesses tiny pores through which air and water is able to pass.

The bark is the wall and roof of a tree. Anyone who has made firewood knows that a tree dries out when its bark is damaged.

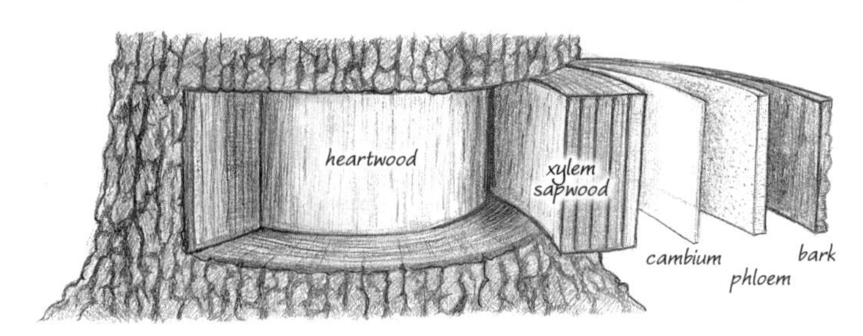

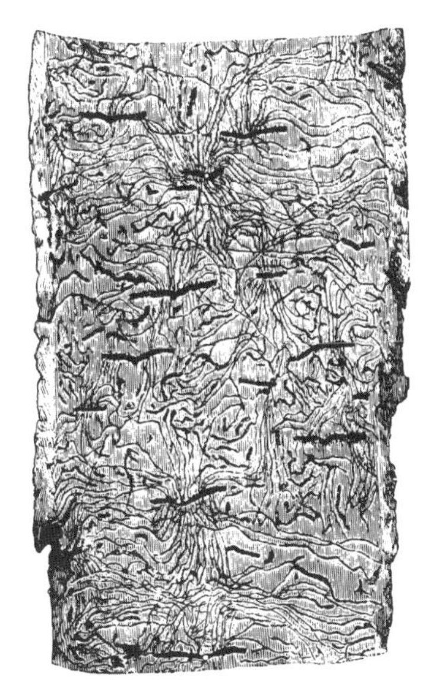

Left: A 'beetle gallery' created by beetles on the inside of a piece of tree bark. Beetles carve into a tree's phloem, or inner bark. Some species carve particular patterns, while others carve away at random.

Facing page: The various layers of a tree, from the heartwood, where energy-rich oils and resins are stored, through the xylem, or sapwood, then the cambrium and phloem, out to the protective bark.

Below: Cork bark harvest, Portugal, c.1870. The cork oak, Quercus suber, is native to southwest Europe and northwest Africa. Unlike removing the bark from most trees, the removal of bark from a cork oak does not hurt the tree, which continues to live, grow and produce new bark, which can be harvested from 9 to 13 years later. Over 15 billion wine bottles are stopped with corks from cork oak trees every year.

The Trunk of a Tree

sapwood, heartwood, and rays

A tree grows from a sapling into one of the majestic beings we see in our forests using its adaptive system of cells to raise water and nutrients in its inner xylem and transport sugars in its outer phloem. It defends itself against damage and parasites using saps delivered via resin canals. A tree's old and dying cells are then reused as its woody backbone, the base from which to rise and reach for the sun.

Each growing season a young tree adds a whole new layer to itself (*opposite lower right*), so grows in thickness and height. As the trunk slowly thickens the woody xylem consists of an inch of new light-colored living *sapwood*, which carries the water and nutrients up through its network of tubes, and inner, older, stronger, darker-colored *heartwood*, whose vessels are clogged with resins, gums, oils, and tannins. Another important feature of trees are its *rays*. Like branches within the wood itself, rays are ribbons of living tissue which run from the center of the tree, right through the cambium and into the phloem. They transport food for storage in the center of the tree, and then carry it back out again when needed.

On the very outside of the phloem, the *cork cambium* produces the vital cork cells which become the bark of the tree (*below*).

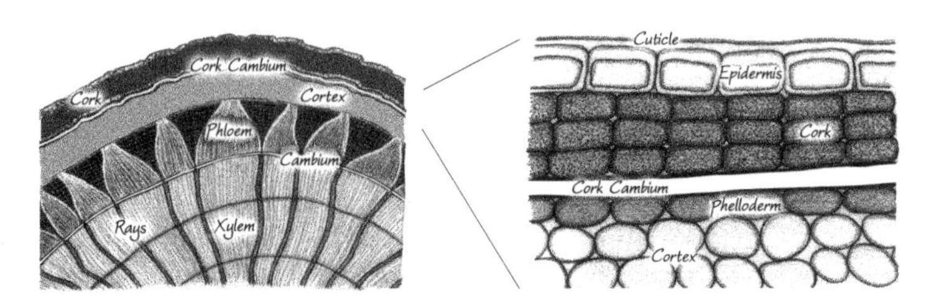

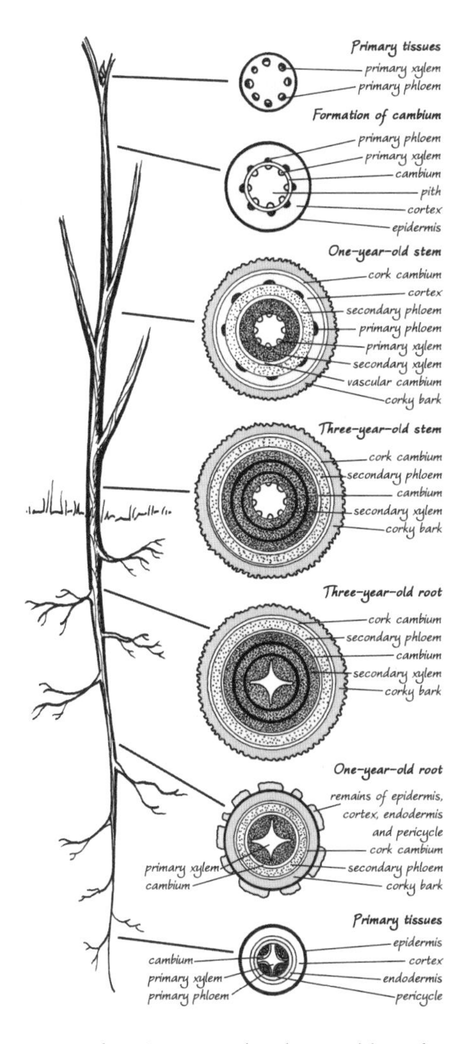

Above: Cross-sections through roots and shoots of a three-year-old tree (after Peter Thomas).

Above: Living rays connect the inner xylem and phloem, transporting food, resins, oils, and other products to and from the central stores.

Above: Every growing season a tree adds a new layer to itself, visible in its annual rings.

THE ROOTS OF A TREE

rock and anchor

A tree's roots are just as important as its foliage, and provide the nutrients and water that allow it to prosper. They also form the anchor upon which the trunk elevates the branches and foliage in the eternal fight for light. Near the trunk, roots can be as wide as railway tracks, and trees recognize the direction of prevailing winds, growing stronger roots in the appropriate direction. Roots further from the trunk are more circular, allowing for strong resistance while providing the optimal carrying capacity. Root ends tend to be near the surface because they need an oxygen-rich environment to grow and the mineralization of nutrients takes place here. Trees only absorb nutrients that have been dissolved in water and they do this by root hairs at the root ends which greatly multiply the contact surface area. Water absorbed via the roots then creates a root pressure that pushes this water slowly upward.

The roots of many trees also live in close symbiosis with fungi in the soil, which are effective decomposers of dead plant material and can even extract minerals from barren rocky soil. Wherever non-decomposed plant material and running water can be found, there are also fungi, and in the diversity of fungi species, there is a lifelong partner for every tree.

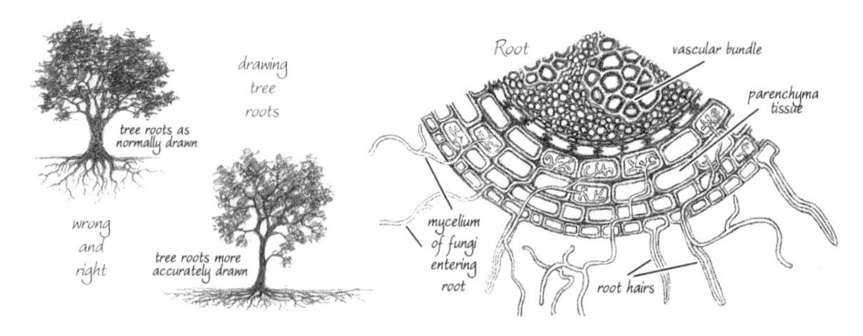

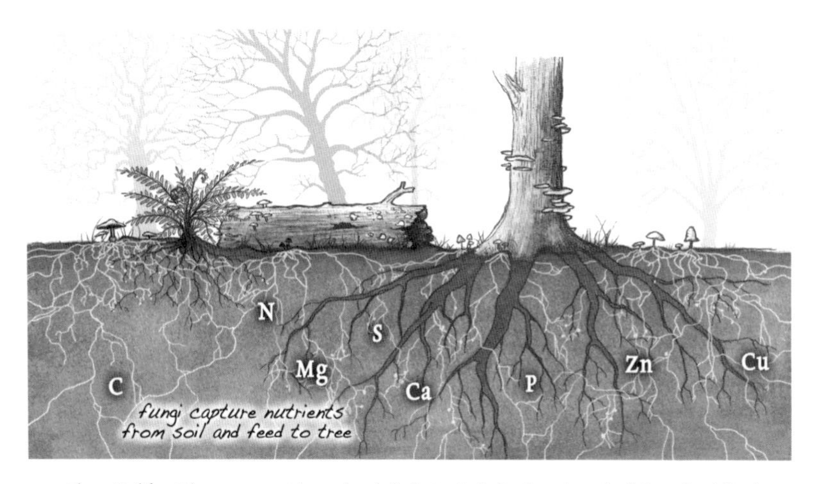

Above: Soil fungi decompose proteins and carbohydrates, including keratin and cellulose. Specialized fungi also form symbiotic mycorrhizal relationships with trees, which receive nutrients from the fungi in exchange for sugars. See diagram on the facing page, showing how fungi mycelium penetrate tree roots.

What Do Trees Eat?

and how do they drink?

In order to grow in height and strength, a tree needs not only energy and (sugar-based) carbohydrates, but also nitrogen and inorganic nutrients (particularly phosphorus). When the supply in the soil is limited, trees obtain these nutrients via mycorrhizal symbiosis, where specialized fungi infect the roots of a tree to obtain carbohydrates in return for delivering these nutrients (*below*).

In the autumn, old foliage falls and is decomposed by microbes into nutrients which dissolve and are retransferred by osmosis to roots. Inside vascular bundles, water and nutrients travel upward through *tracheids*, channels in the new woody xylem of the previous year's annual rings. *Cohesion* ties water molecules into long and durable "strings" inside the pores, while *transpiration* from air pores (stomata) in the leaves removes single water molecules from the top of these strings, causing them to notch upward. Slowly, molecule by molecule, water climbs to the highest leaves of the tallest canopies. A large tree can transpire over 1,000 pints or pounds of water per day.

The sugary output of photosynthesis is also transported through the stem and the branches to all other cells of the tree, but on the *outer* side of the cambium, so just beneath the bark, via sieve tubes. It can move down to the roots or upward toward flowers and developing fruits at speeds of up to 6 feet per hour.

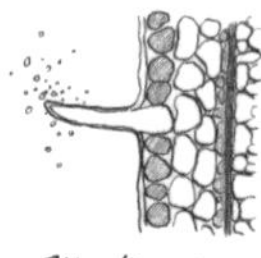
Rhizobium spores infect root hair

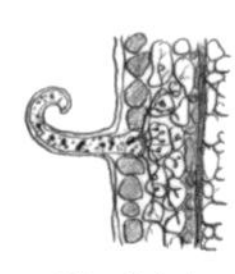
The Nodule begins to develop

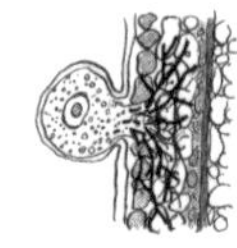
Nodule delivers nitrogen to legume root

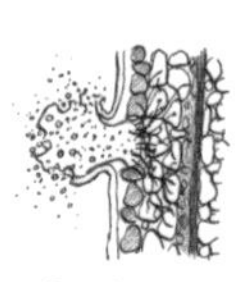
Cocci spores released into soil

Above: All trees need water, nitrogen and nutrients. Mangroves, trees and shrubs which live along shores and rivers, can famously tolerate salty water. Below: The Nitrogen Cycle, upon which all life depends.

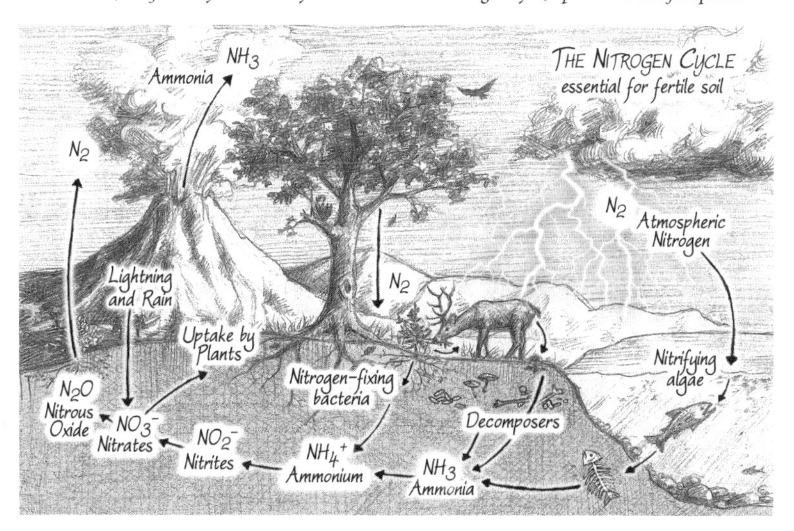

Newton's Apple

the power of gravity

All parts of a tree recognize gravity. Those above the ground strive toward the sun, while underground roots grow downward and cannot live in light. Starches assist with this, by sinking in the cell, and reacting with the cell membrane. Every part of the root and stem of a tree also has *polarity*—any piece of cut tree always grows stem shoots from the top and root shoots from the bottom.

The thickness and length of the stem, branches and roots are all affected by the growth hormone *auxin*. Thus root ends grow close to the surface when sufficient water or nutrients cannot be found deeper, and foliage grows sideways if sufficient space and light exists.

Above and facing page: Rainforest and jungle trees often lean over waterways, stabilizing river banks. They stabilise themselves using the plant growth hormone auxin. In stems, auxin always moves to the darker side of the plant, causing cells there to grow larger than the cells on the lighter side. This makes trunks and branches pull away from the dark ground and curve up to seek the light. In roots, however, auxin inhibits growth, so that shaded cells grow less, causing the root to bend away from the light. Gravity also plays a part, drawing auxin down to the bottom of a branch, so it is in fact a combination of shade and gravity which determine the concentration of auxin.

Left: Isaac Newton observing the famous apple which inspired his theory of gravity. The growth hormone auxin flows downhill, just like rivers flow to the sea, and apples fall from a tree.

Can a Tree Feel Pain?

the plant telegraph system

A rabbit munches the leaves of a young tree on one side of a forest; information about the attack is carried to saplings on the other side of the forest through a network of interconnected roots; trees throughout the area soon begin to secrete chemicals that rabbits find repulsive. To signal "pain" and initiate their defenses, many plants use jasmonic acid, a plant hormone.

Another chemical produced by plant cells is ethylene (C_2H_4). Used in plant metabolism, ethylene is a sweet-smelling, colorless, flammable gas that moves between the cells of a tree. However, ethylene also evaporates easily and can move to other plants in the vicinity. When a tree is under duress or lacks water, it excretes more ethylene, and when the concentration of ethylene surpasses 0.1 parts per million in the air, plants stop growing. Ethylene may also control and accelerate winter preparation in plants. Some trees have other chemicals which have been used for millennia to kill pain in humans. Willow bark contains a substance commonly known as aspirin.

The leaves and needles of a tree are always under constant attack. They get damaged by rain, snow and ice, defoliated by the wind, and attacked by parasites. Their suffering, however is not in vain, for they are shed, decomposed by bacteria and fungi in the soil, and then recyled for the tree to reuse once more as new foliage.

After a tree dies, it is entirely recycled. Nothing is left behind. Fungi and bacteria in the ground recycle the tree into mineral salts and ammonium. Water then carries the recycled products back into the metabolism of other trees. Potassium can cycle back and forth between the soil and the tree up to three times a growth season.

Above left: Mahagony felling, Brazil, c.1890. Cuban Mahogany was logged nearly to extinction across the Caribbean in the early 20th century. Above right: Giant bamboo plantation, Southeast Asia, c.1880.

Is a Tree Greedy?

the battle for resources

Wherever it lives, a tree creates an ecological niche for its own success and, in its battle for survival, must defeat others by winning light, nutrients, and water. A spruce forest, for example, leaves nothing for others and excludes all other plant life; the dispersal of the ground soil and its mineralization is reduced almost to a complete stop. Under the eternal darkness of the canopy, one finds a cellar-like atmosphere where animals and plants cannot survive.

A similar battle for resources occurs underground. The roots of different trees form an underground network, overlapping each other and sometimes growing together. When a root finds a good spot to grow in, it becomes stronger and more effective, while less successful roots slowly wither away.

Within a forest, different trees protect and support each other from the power of the wind by forcing air streams to rise above the canopy. At the edge of the forest, this mutual support system is less effective and trees are much more vulnerable. Trees may not feel pity, and by human standards could be considered greedy; but in the eternal battle for survival, trees are one of the greatest winners.

THE CYCLES OF A TREE

flower summer seed winter

Nature has given trees the ability to survive extreme weather conditions, from subzero temperatures to dry scorching deserts. In the polar extremes below the tree line (where trees can live) in Alaska, Canada, and Siberia, the ground can be frozen up to 27ft deep with only the first 3ft ever defrosting. In the dry Mojave desert, by contrast, a 30ft 1000-year old Joshua tree may have deep roots reaching up to 40ft away, acting like a small oasis.

In cold climates most trees begin photosynthesis as early as possible, although being too hasty can be deadly, as a summer frost will kill light-sensitive cells. Trees in temperate climates also need to maintain their life functions as late into the autumn as possible while making sure that the entire tree goes into winter dormancy. They need to keep water out of their dead cells, or when it freezes it will expand and damage them. Deciduous trees have solved this dilemma by ceasing all activity in the leaves and then shedding them along with harmful waste products in the autumn, while storing energy for flowering and leaf production the following spring.

In tropical rainforests, however, many trees keep their leaves all year round, replacing individuals when they are damaged or worn out, and produce flowers, fruit, and seeds all year round.

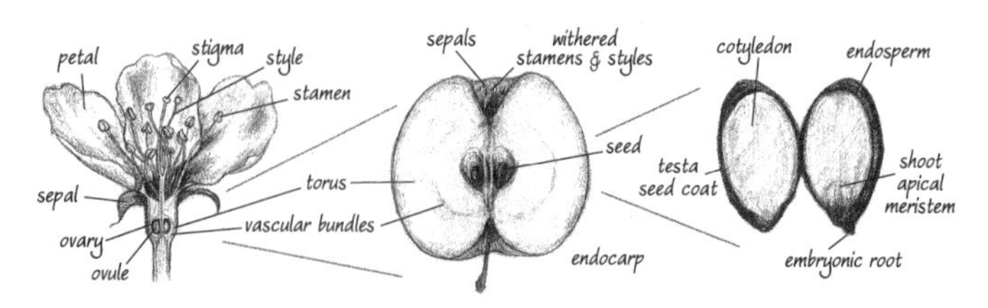

Above: Deciduous trees shed their leaves every autumn, recycling the nutrients back to the soil below, before growing fresh flowers and leaves every spring. Below and opposite: The sex life of a tree. Male sperm cells form in pollen grains in flowers' anthers. Female egg cells form in ovules in flowers' ovaries.

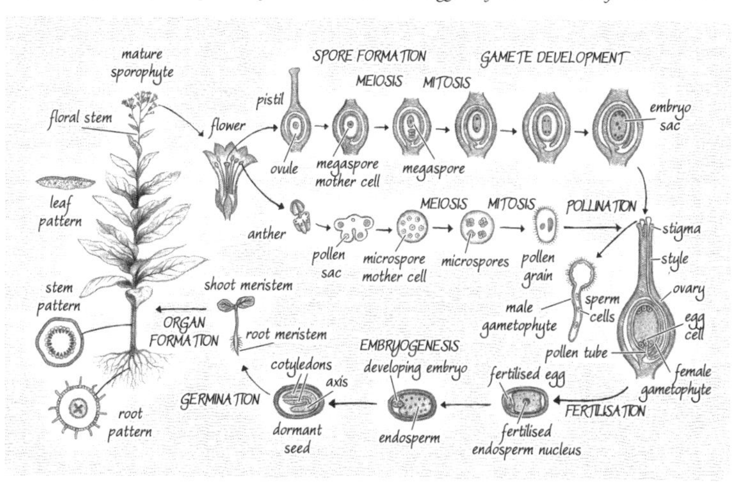

TREE REPRODUCTION

pollen, fruits, and nuts

When a tree matures and has conquered its place in the forest, it begins to devote itself to reproduction. Tree sex is a complicated affair. Some, like willows, aspens, junipers, and poplars are *dioecious* (with separate male and female trees), while others, such as the European ash, completely change sex every year or change the sex of specific branches. Many are hermaphrodites, sporting *perfect* flowers (with male and females parts together), or instead represent the two sexes in separate flowers on the same tree.

Pollination, too, creates particular problems for trees, mostly due to their immense size and sheer output of pollen (birch and hazel produce over 4 million grains per catkin). Conifers and trees in high latitudes tend to be wind-pollinated, while tropical trees mostly employ color, shape, and smell to attract insects, birds, and bats, some highly specialized. To avoid self-pollination trees often develop their male and female parts at different times or heights.

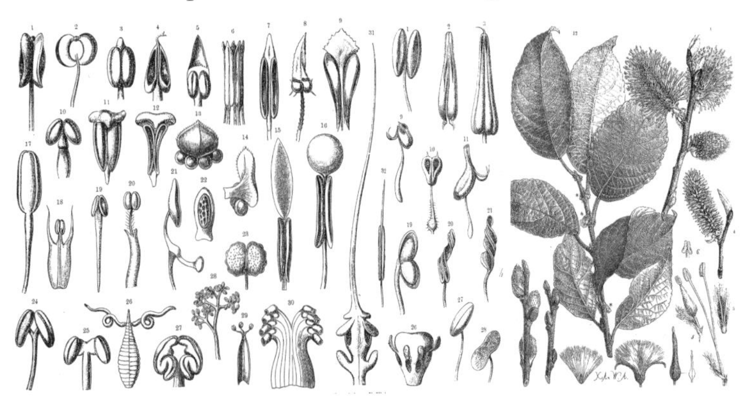

Above: Fig and date palm fruits. Fruits enclose and protect the seeds, and also encourage their dispersal. The fruit, or pericarp, consists of three layers: the skin (exocarp), flesh (mesocarp), and inner part (endocarp). The endocarp can be hard and woody, or fleshy and juicy. Seeds can also be dispersed by animals, wind, and water.

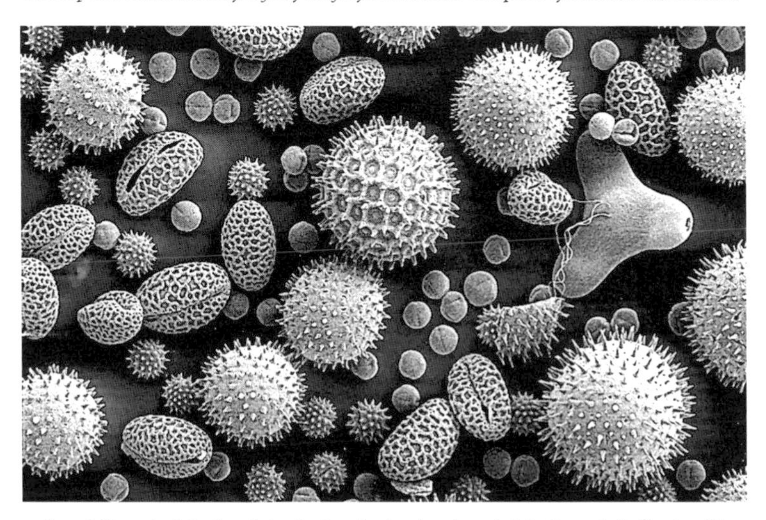

Above: Pollen grains. In hardwoods, it only takes a few days from the arrival of pollen to the fertilization of the ovules. In conifers, however, where the pollen adheres to a sticky pollination drop, it can take longer. Seeds develop at different rates: Elms shed their seeds only 9 weeks after flowering, while pines and red oaks can take 2 years.

Can a Tree See?

the fine art of telling the time

Even the most brutal cold will not kill a forest. A tree in winter dormancy, even in its deepest sleep, remains alive. Yet, to do this, it needs to wake up, year after year, grow, and then fall back asleep at exactly the right time. So how does it keep track of time?

Every tree cell that takes part in photosynthesis contains molecules of *phytochrome*. Even though these account for a mere millionth of the weight of each cell, they determine the prevalent conditions of light with astronomical precision, and allow cells to act and react at precisely the right time. Phytochrome is one of the true miracles of life. A large protein sensitive to the quality, quantity, and direction of red (640 nanometers) and far-red (724 nm) wavelengths of light, it can recognize the length of a day and slow down or accelerate cell processes. Phytochrome affects plants from birth, giving them permission to grow, and stopping them from growing at the wrong time. Only after it has granted permission, can temperature-dependent biochemical activities such as the secretion of enzymes and hormones begin. The activation of the phytochrome is dependent only on light, so can be triggered at any temperature.

Other plant photoreceptors include *cryptochromes* and *phototropins* which are sensitive to light in the blue and ultraviolet wavelengths. All photoreceptors also vitally mediate *phototropism*, where plants grow toward a light source. Here, cells farthest from the light have different concentrations of organ-shaping plant hormones called auxins, resulting in elongated cells on that side, and the plant therefore growing toward the light. A few plants, like vines, exhibit negative phototropism, and grow toward dark solid objects, like trees.

Do Trees Sleep?

when to work and when to rest

In Northern climates, coniferous trees stop growing around the beginning of July. But why so early? It turns out that many trees know when it is time to start preparing for winter, and for sleep.

As nights become longer and the frequency of light changes, information about diminishing sunlight reaches the phytochrome. The tree accordingly begins to prepare resources for the production of the annual shoot of seeds, fruits, and final work on the buds for the next growth season. This process is orchestrated by specialized hormones and enzymes, of which botanists now estimate there are thousands in trees, each specializing in a specific cell function.

During this important time, growth does continue, but as days shorten, so too does the energy received from light (some of which is always also needed for breathing). The oxidation of carbohydrates now secures enough energy to maintain necessary cell functions and the preparations for the upcoming winter. Energy is also needed to remove and store the useful nutrients from old needles and leaves. Abscisic acid, a special hormone used to prepare a tree for winter dormancy, is now secreted (an increased amount of abscisic acid is a sign of reduced capability for photosynthesis). The same process can also happen during dry summer months when the absence of water hinders photosynthesis. With the aid of this acid, a tree begins the process we perceive as leaves changing color, turning yellow.

Winter dormancy is more complex than animal hibernation, though trees, like animals, can also be put to sleep with chloroform and ether, and the same substances that deactivate parts of the human nervous system also work on trees, preventing seed germination.

Above: Winter trees, wood engraving by Gwen Raverat. Below: A tree clock for northern latitudes, showing the important processes that occur over the year. Months are marked as J F M A M J J etc.

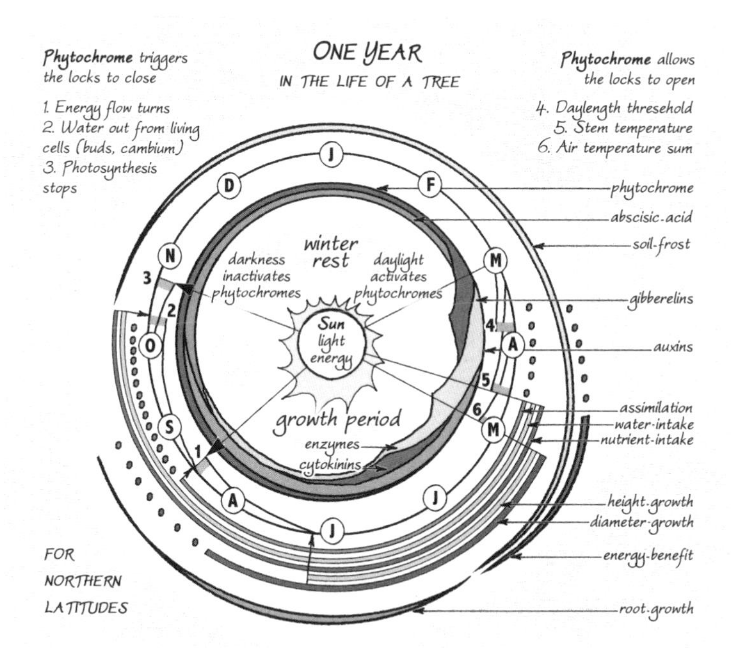

MICROCLIMATES
and other animals

Trees offer animals thousands of opportunities. A tree trunk can appear a hundred times bigger to a small beetle than to a large beetle, and this allows all kinds of tiny beasties to find their own ecological niche. Within cracks in the bark, unique microclimates host exotic combinations of algae and fungi which provide food for insects of many sizes. Under the feet of tiny beetles, further hairline cracks and spots of algae and fungi indicate ecological niches of even smaller insects, like species of bark ticks just 1mm long. At this scale many feed on dust particles that get stuck between the rhizoid of algae and fungi (many birds eat these miniscule life forms). Pine horntail wasps lay their eggs inside the protective bark of old or damaged trees, their offspring later feeding on the tree from the inside. Woodpeckers, in turn, dig these parasites out!

Sometimes, stems of trees can grow quicker than the bark, causing cracking (also caused by wind). Spruce trees try to repair such damage by insulating the cracks with sap, but wood-destroying fungi or polypores often get in and start infiltrating mycelium, which in turn is eaten by a wide range of insects. Many trees, therefore, are already eaten before they tumble over. Only the durable bark keeps the rotten tree erect long after it has lost the battle to parasites.

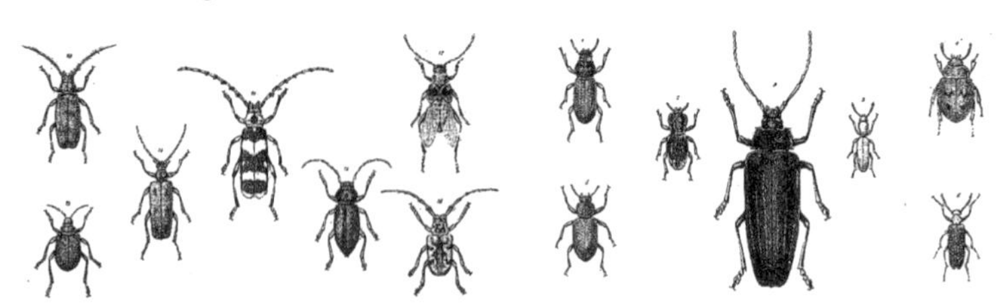

Above left and below: Moss and lichen: Above right: South American tree-climbing cheese plant. Facing page: Beetles thrive on tree bark. A tree produces and stores huge quantities of different materials and, in the often fast-moving cycles of nature, is a relatively stable player. As a result, bark and moss provide food and shelter for a multitude of species. After doing their jobs, flowers and seeds provide crucial nutrition for others. The willow ptarmigan and black grouse eat the nutrition-rich buds of trees; squirrels are specialized in seeds, while both the rabbit and the moose eat the bark of an aspen (during summer, moose enjoy the leaves of young saplings as well as fresh shoots). Small animals are everywhere more abundant and diverse than bigger ones. The constantly-changing micro-structures in trees, and the rainwater which flows down their branches, form gardens of food. This confirms the old adage that where there is flowing water there is also life.

Trees and Soil

the air we breathe

As the biomass of an area increases, more and more nutrients are tied into trees. In tropical rainforests (*opposite*), despite the recycling and decomposition of tree litter by different fungi, almost all the nutrients are immediately transported back into the canopies above, leaving the shady soil nearly sterile. By contrast, in the boreal climate of snow and frost, decomposing is so slow that as the forest grows older the soil becomes thicker, the moss layer and ground debris slowing decomposition almost to a stop. Here, mineralized nutrients in the soil diminish as the trees become bigger and older, resulting in the minimization of all forms of life in old forests. In climates in between, in temperate forests, most of the nutrients exist in the soil.

Plants make soil and many tree species improve soil by tying nitrogen from the air with the help of bacteria in their roots, also benefitting other nearby trees. Forest fires, too, can improve the soil for the next generation. The soil of conifer forests often becomes acidic over time, and the ashes created in a forest fire neutralize this. In fires, all plant materials are mineralized back into nutrients.

Trees also improve the air. In the boreal climates, fast-growing forests are important consumers of carbon dioxide, which they bind into their biomass. Tropical rainforests, by contrast, produce as much carbon dioxide as they consume. All trees, however, manufacture oxygen for us to breathe. We can feel this when walking in the forest, especially during calm nights when oxygen excretion is high and the wind does not disperse it quickly. The large surface of a tree canopy is also an effective air-filter. Inside a thick forest, there can be ten times fewer dust particles than in an open area.

Musical Trees

the song of the forest

In nature, we hear the song of trees the loudest when a woodpecker is drumming, announcing its search for a mate. Humans too have always found trees well-suited for musical instruments, as wood provides the best acoustical building material. Good wood can amplify and reproduce sound again and again exactly as a player wishes.

For the expert manufacture of instruments, only branchless, straight and non-spiraled trees are chosen, particularly those that have grown evenly for over a hundred years in dense forests, with narrow annual rings. Instrument-makers select different species of tree for different puposes; for example, the cover of a violin is always built from strong-sounding spruce, its sides from delicate maple, while xylophones are traditionally made from rosewood.

The cell structure of wood is such that it does not shrink or expand, so vibrations resonate strongly, amplifying any sound. It is also miraculously durable—Stradivarius violins, over 300 years old, still today increase in value the more they are played. Thus in the ever-changing songs of trees, wood becomes a refined musical instrument, an invaluable treasure for creating new beautiful music.

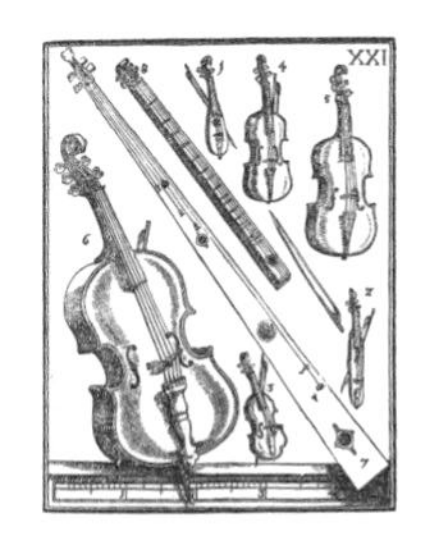

Left: Bamboo plants tower over a river in Southeast Asia. The natural hollow tubes of bamboos make them ideal for fashioning various musical instruments. Bamboo can be used to create flutes, trumpets, rattles, drums, xylophones, didgeridoos, guitars, panpipes and mouth organs. The most famous flute in India, played by Lord Krishna, is traditionally made of bamboo.

Below left: Coconuts shells are hugely versatile. Their nearly spherical form means they are ideal for maracas and other shakers. Cut in half, they can be used to make kalimbas, hand-drums, and small-bowled string instruments like the Chinese Huqin or Iranian Kamanche.

Below right: Austrian spruce, for hundreds of years the tree of choice for violin-makers.

TREES, HUMANS, AND ENERGY

learning respect

From the spikes in a rake, to the rim of a wheel, trees have long provided the central raw material for human survival. A tree is the most durable, constantly growing and self-repairing store of resources we know. Wood can be sawed and carved, nailed, glued, pressed, and heated. It can be chemically broken down into cellulose, lingine and other compounds, paper, energy, foods, lubricants, medicines and textiles, often producing spare energy in the processes.

Forest industries have always been self-sufficient and cost-effective. The leftovers from the saw industries are a source of fuel and raw materials for the pulp industries. In countries where forest resources are abundant, energy production based on trees can be decentralized as raw materials are easily available locally.

Trees store the products of each growth season into the cells and chemicals of their stems, branches, and roots, protecting these warehouses from parasites using resinous saps and other chemicals. As a tree grows, therefore, it contains more and more energy, and since the needles of coniferous trees contain waxes and other oils, one cubic meter of tree can equal 220 liters of oil in energy. Indeed, one hectare of full-grown forest can contain as much energy as 60,000 liters of oil. Among other uses, this can be used to generate large amounts of electricity. So a forest is like a charged battery, a battery that can be worth much more per hectare than its raw material.

Throughout history, we have always used trees for energy. Whether firewood to keep us warm and cook food, or today the known fossil fuels, peat, coal, natural gas and oil, all formed from the rich biomass of forests, slowly compressed beneath layers of earth.

A tree's life and death also gives us the best growth elixir known to man: tree ash. Adding ash to the ground miraculously increases fungi activity a thousandfold. This can result in nearly instantaneous improvements in the soil and therefore in the health and growth of a tree, lasting tens of years into the future. If we compare this, for example, with nuclear energy, we see how primitive nuclear energy is. Producing energy with nuclear power leaves behind heaps of waste which destroy everything in the immediate environment for thousands of years. A tree, on the contrary, recycles everything.

Climate Change

heads up, here we go

Carbon dioxide is the first and most abundant of the "greenhouse gases" (methane is the second). More simple molecules in the atmosphere, of say nitrogen or oxygen, are not long enough to vibrate in the infrared and trap heat. This is why small increases in the amount of CO_2 in the atmosphere, or tiny additions of methane, can have enormous and rapid repercussions for terrestrial temperatures.

The climate has changed before and it will change again. Indeed, some trees benefit from increased carbon dioxide. The 4°C rise in temperature that Northern Europe experienced at the end of the last ice age merely caused deciduous trees to flourish in more northern climates and coniferous trees, in turn, to slowly move northward. This time, however, the climate is changing much faster than it ever has before, affecting the whole planet. We must therefore understand the miracle of trees, so that we can understand their life and death, and so be able to prevent what threatens them, and, with them, all of us.

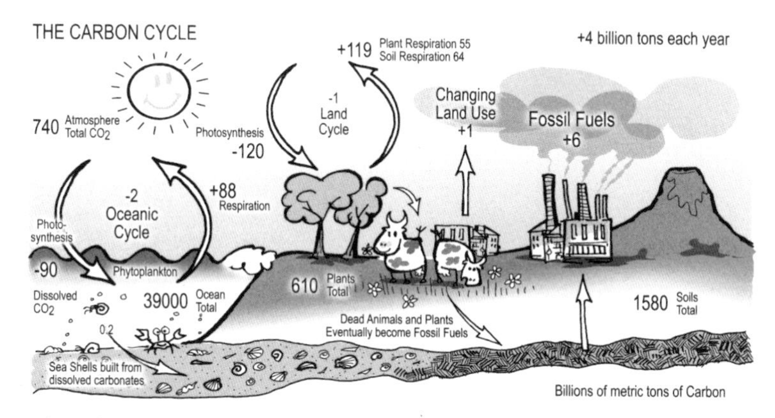

Above: The Carbon Cycle. Trees are the best device ever invented for removing carbon from the atmosphere. As long as humans continue to cut them down and replace them with cash crops, our future on this planet is uncertain. Below: The Cowthorpe Oak, near Wetherby, Yorkshire, was once the largest and oldest oak in England. It fell in 1950 after being struck by lightning. Facing page: On Monday 27th June 1853, a huge 1244-year-old giant sequoia was felled in Calaveras, California, after which a ball was held on the stump.

Select Plant Orders with Trees

NB - *See full diagram on page 9, where dots represent the trees in the orders listed below. The often thousands of species of other smaller plants in many of these orders are not shown here, and neither are orders without significant trees.*

SEEDLESS PLANTS

PTEROPHYTES - *tree ferns, and (extinct) giant horsetails.*

PLANTS WITH SEEDS

GYMNOSPERMS *(without flowers)*

CYCADALES - *cycads, the most ancient trees still with us, predate the dinosaurs, dioecious and palm-like.*

GINKGOALES - *second most ancient survivor. Just one species survives, Ginkgo biloba.*

CONIFERAE

PINACEAE - *pines, cedars, firs, spruces, larches, hemlocks.*

CUPRESSALES - *cypresses, thujas, junipers, redwoods - yews.*

ARAUCARIALES - *monkey-puzzles - rimu, totara, kahikatea - celery pines.*

GNETALES - *ephedras, jointfirs, gnetums.*

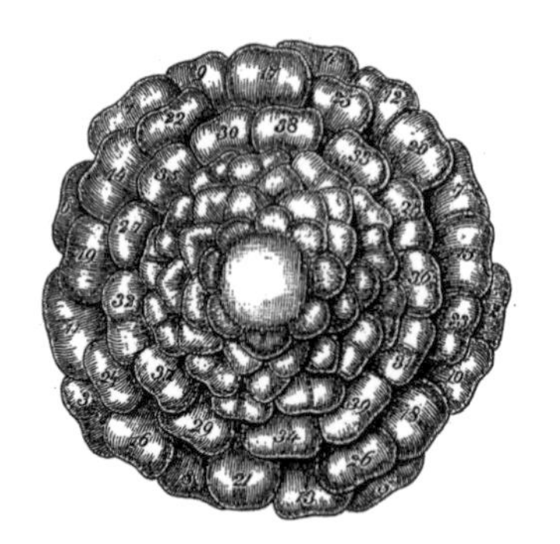

ANGIOSPERMS *(with flowers)*

MAGNOLIIDS

CANELLALES - *winter's bark trees, white cinnamons.*

LAURALES - *greenhearts, stinkwoods, Queensland "walnuts," avocados, bay trees, sassafrasses, cinnamons, camphors.*

MAGNOLIALES - *magnolias (one of the first flowering trees), tulip trees, custard apples, nutmegs.*

MONOCOTS: COMMELINIDS

POALES - *pineapples, bamboos.*

ZINGIBERALES - *bananas, traveler's palms.*

ARECALES - *palms (over 2,500 species).*

MONOCOTS: LILIOIDS

PANDANALES - *screw pines.*

ASPARAGALES - *butcher's brooms, yuccas, agaves, Joshua trees, dragon trees, aloes.*

EUDICOTS: PRIMITIVE

PROTEALES - *bottlebrush trees, hakeas, fynbos, rewararas, silk oaks, planes, macadamia nuts, box.*

TROCHODENDRALES - *trochodendrons.*

CARYOPHYLLALES - *didiereaceae, cacti.*

SANTALALES - *hog plums, African "walnuts" (not the same as in Sapinales, below), sandalwoods, garlic fruits, mistletoes.*

EUDICOTS: ROSIDS

SAXIFRAGALES - *liquidambars, sweet gums, witch hazels, katsuras.*

ZYGOPHYLLALES - *creosotes, lignum vitae, bulsnias.*

CELASTRALES - *spindle trees, khat trees, kokoons.*

MALPIGHIALES - *poplars, rubber trees, euphorbias, cassavas, candlenut trees, tung trees, Chinese tallows, mangroves, willows, odokos, Barbados cherries, garcinias, mangosteens, ekkis.*

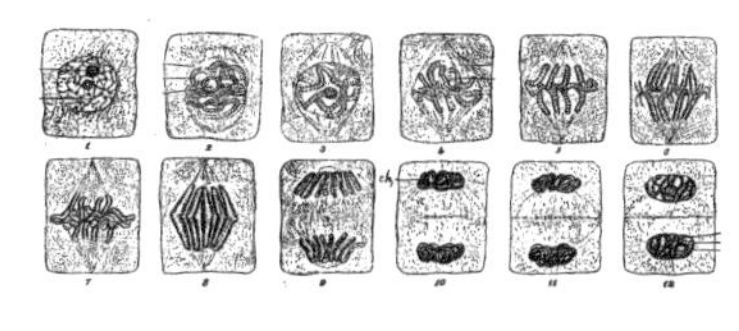

OXALIDALES - *star fruits, coachwoods.*

MYRTALES - *pomegranates, fuchsia trees, Indian laurels, Indian silver-grays, idigbos, afaras, limbas, myrtles, clove trees, cajuputs, allspices, guavas, eucalyptus, gum trees, mountain ash (not rowan).*

MALVALES - *limes (lindens), American basswoods, danta, Indian wild almond, pterygotas, mengkulangs, cocoas, colas, baobabs, kapoks, bombaxes, red silk cotton trees, balsas, durians, dipterocarps, shoreas, mersawas, krabaks.*

BRASSICALES - *moringas.*

SAPINALES - *quassias, picramnias, white syringas, ailanthuses, frankincenses, myrrhs, gaboon, lemons, organges, limes, grapefruits, Queensland maples, southern silver ashs, pau marfins, Ceylon satinwoods, mahoganies, pride of India's, African walnuts, South American cedars, sapeles, neems, sumacs, smoke bushes, pistachios, cashews, mangos, lychees, rambutans, akees, koelreuterias, xanthoceras, taun trees, maples, horse chestnuts, buckeyes.*

EUDICOTS: ROSIDS THAT FIX NITROGEN

FAGALES - *oaks, beeches, sweet chestnuts, birches, alders, hazels, hornbeams (ironwoods), she-oaks, walnuts, hickories, pecans, wingnuts, bayberries, wax myrtles, candelberries.*

FABALES - *acacias, Australian blackwoods, leucaenas, dalbergias, Indian rosewoods, sissoos, African blackwoods, Brazilian tulipwoods, Rhodesian teaks, zebranos, purplehearts, Borneo teaks, angelims, brazilwoods, rosewoods, laburnums, Judas trees, parkias, mimosas, asokas, prosopises, mesquites, sophoras, Japanese pagoda trees, ingas, tamarinds.*

ROSALES - *apples, pears, quinces, apricots, peaches, plums, cherries, hawthorns, blackthorns, firethorns, shadbushes, rowans, whitebeams, buckthorns, ceanothuses, raisin trees, jujubes, elms, zelkovas, hackberries, sugerberries, jackfruits, breadfruits, breadnuts, irokos, snakewoods, mulberries, figs, banyans, peepuls, cecropias.*

EUDICOTS: ASTERIDS

CORNALES - *dogwoods, handkerchief trees, tupelos.*

GARRYALES - *bayberry trees.*

APIALES - *umbrella trees.*

DIPSICALES - *snowberry trees, elders.*

ERICALES - *diospyroses, persimmons, date plums, black sapotes, ebonies, star apples, sapodillas, shea butter trees, chicles, palaquiums, buckthorns, cherry mahogonies, makores, nyotahs, tambalacoque trees, ardisias, Franklin trees, camellias, Australian silk oaks, rhododendrons, strawberry trees, madrones, carinianas, brazil nuts, paradise nuts, lecythises, cannonball trees.*

SOLANALES - *jimsonweeds, solanaceae.*

GARRYALES - *cordias.*

GENTIANALES - *gardenias, coffees, cinchonas, aburas, opepes, degames, dyeras, rosa perobas, frangipanis, fragaeas.*

LAMIALES - *ashs, olives, eremophilas, myoporums, privets, jacarandas, catalpas, sausage trees (kigelias), tulip trees (different from the one in Magnoliales), teaks, gmelinas.*

AQUIFOLIALES - *ilexs, hollies.*

ASTERALES - *muhuhus.*

GLOSSARY

ABSCISIC ACID: *Abscisic acid (ABA), also known as abscisin II and dormin, is a plant hormone. It functions in many plant developmental processes, including abscission and bud dormancy.*

ADP-MOLECULE: *Adenosine diphosphate is a nucleotide ester of pyrophosphoric acid with the nucleoside adenosine. ADP consists of the pyrophosphate group, the pentose sugar ribose, and the nucleobase adenine. ADP is the product of ATP dephosphorylation by ATPases. ADP is converted back to ATP by ATP synthases. ATP is an important energy transfer molecule in cells.*

ANNUAL RINGS: *Growth rings, also referred to as tree rings or annual rings, can be seen in a horizontal cross section cut through the trunk of a tree. Growth rings are the result of new growth in the vascular cambium, a lateral meristem, and are synonymous with secondary growth. Visible rings result from the change in growth speed through the seasons of the year, thus one ring usually marks the passage of one year in the life of the tree.*

ARGININE: *Arginine (abbreviated as Arg or R)[1] is an α-amino acid. The L-form is one of the 20 most common natural amino acids.*

ASPARAGINE: *Asparagine (abbreviated as Asn or N; Asx or B represent either asparagine or aspartic acid) is one of the 20 most common natural amino acids on Earth.*

ATP-MOLECULE: *Adenosine 5'-triphosphate is a multifunctional nucleotide that is most important as a "molecular currency" of intracellular energy transfer.[1] In this role, ATP transports chemical energy within cells for metabolism. It is produced as an energy source during the processes of photosynthesis and cellular respiration and consumed by many enzymes and a multitude of cellular processes including biosynthetic reactions, motility and cell division.*

AUXIN: *A class of plant growth substance (often called phytohormone or plant hormone). Auxins play an essential role in coordination of many growth and behavioral processes in the plant life cycle.*

BARK: *Bark, also known as periderm, is the outermost layer of stems and roots of woody plants such as trees. It overlays the wood and consists of three layers, the cork, the phloem and the vascular cambium.*

CAMBIUM: *is a layer or layers of tissue, also known as lateral meristems, that are the source of cells for secondary growth.*

CELLOPHANE: *A thin, transparent sheet of regenerated cellulose.*

CELLULOSE: *The primary structural component of the cell walls of green plants. It is the most common organic compound on Earth, about 33% of all plant matter. The cellulose content of cotton is 90%, while wood is about 45% cellulose.*

CHLOROPHYLL: *A green pigment found in most plants, algae, and cyanobacteria. Chlorophyll absorbs light most strongly in the blue and red but poorly in the green portions of the electromagnetic spectrum, hence the green color of chlorophyll-containing tissues like plant leaves.*

CHLOROPLASTS: *Organelles found in plant cells and eukaryotic algae that conduct photosynthesis. Chloroplasts absorb light and use it in conjunction with water and carbon dioxide to produce sugars, the raw material for energy and biomass production in all green plants and the animals that depend on them, directly or indirectly, for food.*

DECOMPOSERS: *Decomposers (or saprotrophs) are organisms that consume dead organisms, and, in doing so, carry out the natural process of decomposition. Decomposers use deceased organisms and non-living organic compounds as their food source. The primary decomposers are bacteria and fungi.*

EGG CELL: *A haploid female reproductive cell or gamete. The term ovule is used for the young ovum of an animal, as well as the plant structure that carries the female gametophyte and egg cell and develops into a seed after fertilization.*

ELECTROMAGNETIC WAVES: *Electromagnetic (EM) radiation (also called light even though it is not always visible) is a self-propagating wave in space with electric and magnetic components. Electromagnetic radiation is classified into types according to the frequency of the wave: these types include, in order of increasing frequency, radio waves, microwaves, terahertz radiation, infrared radiation, visible light, ultraviolet radiation, X-rays and gamma rays.*

EMBRYO: *A multicellular diploid eukaryote in its earliest stage of development, from the time of first cell division until birth, hatching, or germination.*

EPIDERMIS: *The outer single-layered group of cells covering a plant, especially the leaf and young tissues of a vascular plant including stems and roots. Epidermis and periderm are the dermal tissues in vascular plants. The epidermis forms the boundary between the plant and the external world. The epidermis serves several functions: protection against water loss, regulation of gas exchange, secretion of metabolic compounds, and (especially in roots) absorption of water and mineral nutrients.*

EPIDERMIC CELLS: *see epidermis.*

ETHYLENE: *Ethylene (or IUPAC name ethene) is the chemical compound with the formula C_2H_4. It is the simplest alkene. Because it contains a double bond, ethylene is called an unsaturated hydrocarbon or an olefin. It is extremely important in industry* and even has a role in biology as a hormone. Ethylene is the most produced organic compound in the world; global production of ethylene exceeded 75 million metric tonnes per year in 2005.

FORCE OF COHESION: Cohesion (v. lat. cohaerere "stick or stay together") or cohesive attraction or cohesive force is a physical property of a subtsance, caused by the intermolecular attraction between like-molecules within a body or substance that acts to unite them.

FUNGI: A fungus is any eukaryotic organism that is a member of the kingdom Fungi. The fungi are heterotrophic organisms possessing a chitinous cell wall. The majority of species grow as multicellular filaments called hyphae forming a mycelium; some fungal species also grow as single cells.

GLUTAMINE: Glutamine (abbreviated as Gln or Q; the abbreviation Glx or Z represents either glutamine or glutamic acid) is one of the 20 amino acids encoded by the standard genetic code. Its side chain is an amide formed by replacing the side-chain hydroxyl of glutamic acid with an amine functional group.

INNER MEMBRANE: The inner membrane is the biological membrane (phospholipid bilayer) of an organelle or Gram-negative bacteria that is within an outer membrane.

JASMONIC ACID: Jasmonic acid (JA) is a member of the jasmonate class of plant hormones. The major functions of JA in regulating plant growth include growth inhibition, senescence, and leaf abscission.

MERISTEMATIC CELLS: A meristem is a tissue in all plants consisting of undifferentiated cells (meristematic cells) and found in zones of the plant where growth can take place.

MYCELIUM: Mycelium is the vegetative part of a fungus, consisting of a mass of branching, thread-like hyphae. Fungal colonies composed of mycelia are found in soil and on or in many other substrates.

MYCORRHIZAL SYMBIOSIS: fungus roots coined by Frank, 1885[1]; typically seen in the plural forms mycorrhizae or mycorrhizas) is a symbiotic (occasionally weakly pathogenic) association between a fungus and the roots of a plant.[2] In a mycorrhizal association the fungus may colonize the roots of a host plant either intracellularly or extracellularly.

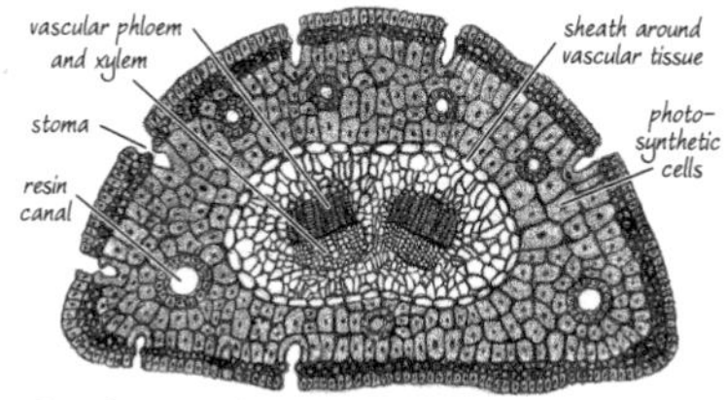

Cross-Section of a Pine Needle

OSMOSIS: Osmosis is the diffusion of water through a cell wall or membrane or any partially-permeable barrier from a solution of low solute concentration to a solution with high solute concentration, up a solute concentration gradient.

OUTER MEMBRANE: The outer membrane refers to the outside membranes of Gram-negative bacteria, the chloroplast, or the mitochondria. It is used to maintain the shape of the organelle contained within its structure, and it acts as a barrier against certain dangers.

PARENCHYMA CELLS: are thin-walled cells of the ground tissue that make up the bulk of most nonwoody structures, yet sometimes their cell walls can be lignified. Parenchyma cells in between the epidermis and pericycle in a root or shoot constitute the cortex, and are used for storage of food.

PHLOEM: In vascular plants, phloem is the living tissue that carries organic nutrients (known as photosynthate), particularly sucrose, a sugar, to all parts of the plant where needed. In trees, the phloem is the innermost layer of the bark. The phloem is mainly concerned with the transport of organic material, such as glucose and starch, made during photosynthesis.

PHOSPHATE: In inorganic chemistry, is a salt of phosphoric acid.

PHOTONS: Photon is the elementary particle responsible for electromagnetic phenomena. It is the carrier of electromagnetic radiation of all wavelengths, including gamma rays, X-rays, ultraviolet light, visible light, infrared light, microwaves, and radio waves.

PHYTOCHROME: Phytochrome is a photoreceptor, a pigment that plants use to detect light. It is sensitive to light in the red and far-red region of the visible spectrum. Many flowering plants use it to regulate the time of flowering based on the length of day and night (photoperiodism) and to set circadian rhythms. It also regulates other responses including the germination of seeds, elongation of seedlings, the size, shape, and number of leaves, the synthesis of chlorophyll, and the straightening of the epicotyl or hypocotyl hook of dicot seedlings.

PHOTOSYNTHESIS: Photosynthesis is the conversion of light energy into chemical energy by living organisms. The raw materials are carbon dioxide and water; the energy source is sunlight; and the end-products are oxygen and (energy rich) carbohydrates, for example sucrose, glucose, and starch. This process is arguably the most important biochemical pathway, since nearly all life on Earth either directly or indirectly depends on it.

PIGMENTS: A pigment is a material that changes the color of light it reflects as the result of selective color absorption.

POLARITY: Polarity refers to the dipole-dipole intermolecular forces between the slightly positively-charged end of one molecule to the negative end of another or the same molecule.

POLLEN: The pollen grain with its hard coat protects the sperm cells as they move from the stamens of the flower to the pistil of the next flower.

RESIN: *Resin or rosin is a hydrocarbon secretion of many plants, particularly coniferous trees.*

ROOT HAIRS: *Root hairs, the rhizoids of many vascular plants, are tubular outgrowths of trichoblasts, the hair-forming cells on the epidermis of a plant root. That is, root hairs are lateral extensions of a single cell and only rarely branched.*

ROOT PRESSURE: *Root pressure is the osmotic pressure within the cells of a root system that causes sap to rise through a plant stem to the leaves.*

SIEVE TUBES: *Sieve vascular tissue tube elements, also called sieve tube members, are a type of elongated parenchyma cells in phloem tissue. The main function of the sieve tube is transport of carbohydrates in the plant (e.g., from the leaves to the fruits and roots).*

STEM: *A stem is one of two main structural axes of a vascular plant.*

STOMATA: *A stoma (also stomate; plural stomata) is a tiny opening or pore, found mostly on the underside of a plant leaf and used for gas exchange. Air containing carbon dioxide enters the plant through these openings where it is used in photosynthesis and respiration.*

SYMBIOSIS: *The term symbiosis commonly describes close and often long-term interactions between different biological species.*

TRACHEID: *Tracheids are elongated cells in the xylem of vascular plants, serving in the transport of water. The build of tracheids will vary according to where they occur. The two major functions that tracheids may fulfill are: as part of the transport system and in structural support.*

TRANSPIRATION: *Transpiration is the evaporation of water from the aerial parts of plants, especially leaves but also stems, flowers, and roots. Leaf transpiration occurs through stomata, and can be thought of as a necessary "cost" associated with the opening of stomata to allow the diffusion of carbon dioxide gas from the air for photosynthesis. Transpiration also cools plants and enables mass flow of mineral nutrients from roots to shoots.*

ULTRAVIOLET RADIATION: *Ultraviolet (UV) light is electromagnetic radiation with a wavelength shorter than that of visible light, but longer than soft X-rays. It is so named because the spectrum consists of electromagnetic waves with frequencies higher than those that humans identify as the color violet (purple).*

VASCULAR BUNDLES: *A vascular bundle is a part of the transport system in vascular plants. The transport itself happens in vascular tissue, which exists in two forms: xylem and phloem. Both these tissues are present in a vascular bundle, which in addition will include supporting and protective tissues.*

WINTER DORMANCY: *Dormancy is a period in an organism's life cycle when growth, development, and (in animals) physical activity is temporarily suspended. This minimizes metabolic activity and therefore helps an organism to conserve energy. Dormancy tends to be closely associated with environmental conditions.*

XYLEM: *In vascular plants, xylem is one of the two types of transport tissue, phloem being the other. Its basic function is to transport water.*